Biotechnological Applications for Improvement of *Solanum surattense* A Medicinal Plant

Biotechnological Applications for Improvement of *Solanum surattense*
A Medicinal Plant

Dr. N. Rama Swamy

2024

Daya Publishing House®
A Division of
Astral International Pvt. Ltd.
New Delhi – 110 002

ISBN : 9789351306375 (HB)

Published by : **Daya Publishing House®**
A Division of
Astral International Pvt. Ltd.
– ISO 9001:2015 Certified Company –
4736/23, Ansari Road, Darya Ganj
New Delhi-110 002
Ph. 011-43549197, 23278134
E-mail: info@astralint.com
Website: www.astralint.com

DEDICATED TO

With humility and reverence to the fond memory of my beloved parents who encouraged and flared passion in me to learn more for ever.

Prof. V Gopal Reddy, ***Ph.D.***
Vice-Chancellor

KAKATIYA UNIVERSITY
Vidyaranyapuri
Warangal - 506 009, A.P.
Phones: 0870-2439966 (O) Fax: 0870-2439600
Email : vc@kakatiya.ac.in; kuvcpeshi@yahoo.co.in

No. 79/VC/KU/2006 18 April, 2006

Foreword

Plant Tissue Culture and Biotechnology plays an important role in Agriculture, Medicine and Industry. There has been lot of development in plant biotechnology for the last one decade. In this book the author Dr. Rama Swamy, Nanna has focused his extensive research work on the species using biotechnological techniques for improvement of Indian Solanum. Researcher will benefit greatly from this clearly presented data with thorough review on the subject.

I congratulate Dr. Swamy for his sincere efforts in bringing out this book. His wide experience in the field of plant biotechnology and exposure to different-International Laboratories/Institutes have certainly helped him to contribute in a form of book. This book by Dr. Swamy will undoubtedly contribute significantly for the benefit of researchers of Plant science, Agriculture, Medicinal Plant Biotechnology and Ayurvedic Medicine.

(Prof. V. Gopal Reddy)

1-8-110, 'Bharathi Nilayam', Balasamudram, Hanamkonda - 506 001 (R) 0870-2455888

Preface

Plant Tissue Culture Technology plays a major role in plant biotechnology. Advances made in genetic engineering and molecular biology can be exploited through the techniques developed in plant tissue culture. Considerable progress has been made to understand the *in vitro* culture of plants including the cell suspension culture and production of Pharmaceutically, Medicinally important compounds and transformation technologies to produce transgenic plants for agricultural improvement. Plant Biotechnology has contributed a lot in developing insect, pest, herbicide and drought tolerant transgenic plants including the production of edible vaccines and therapeutically important proteins.

The present book essentially comprised of 11 Chapters, namely: Introduction (dealing periodical improvement in plant tissue culture and biotechnology), Introduction to Species (especially review on the species), Callus Induction, Plant regeneration (callus mediated and direct), Somatic embryogenesis, Micropropagation (meristem and nodal culture), Floral bud culture, Androgenic haploidy, NaCl, KCl and Mannitol tolerant cell lines, Antibiotic resistance and *Agrobacterium* mediated genetic transformation.

Keeping in view of the interest of researchers in plant biotechnology, each chapter is sufficiently supplemented with illustrations, tables, figures and photographs.

I hope this book, apart from catering the specific requirement of Plant Science Researchers, will also be useful to researchers working in the area of Agriculture, Horticulture, Pharmacy and Ayurvedic Medicine. I have made an attempt to present the subject in lucid manner and have taken all possible care. Still, I invite constructive criticism for its improvement.

I am highly indebted to my parents who have given me strength and knowledge to complete this work.

I am grateful to UGC unit, Kakatiya University and also University Grants Commission, New Delhi for providing the financial assistance in shaping and publishing the book.

I am grateful to my teachers Prof. Bir Bahadur for his inspiration and Prof. K. Subhash for his continuous encouragement and support in taking up this task. I am thankful to my teachers Prof. Vidyavati (former Vice-chancellor), Prof. Digamber Rao and Prof. S.M. Reddy.

I express gratitude to my colleagues Prof. N. Pratap Reddy, Prof. C. Suvartha, Prof. A. Sadanandam, Prof. B. Mallaiah, Prof. M.A. Singaracharya, Prof. K. Jaganmohan Reddy, Prof. S. Ram Reddy, Prof. V.S. Raju, Dr. S. Girisham, Dr. B. Digamber Rao, Dr. A. Raghan, Dr. A. Seetharam, Dr. V. Krishna Reddy, Dr. P. Venkataiah, Dr. A.V. Rao and Dr. T. Christopher for their encouragement. I also acknowledge my research scholars Dr. T. Ugandhar, Dr. A. Laxman, Dr. M. Praveen, Mr. M. Rambabu, Mr. M. Upender, Mr. K. Appa Rao, Mr. M. Ghansingh, Mr. G. Vikram, Mr. P. Mahendar and Mrs. D. Sharada for their help.

I am especially grateful to Prof. V. Gopal Reddy, Ho'ble Vice-Chancellor, Kakatiya University for his foreword and encouragement.

I am thankful to my elders whose continuous support and inspirations led me to complete this work. I am fortunate in having a family, which understands the preoccupation that goes with such projects. I am grateful to my wife Vanaja and to my daughter Shwetha and son Varun Koushik for their cooperation and patience.

Dr. Rama Swamy, Nanna

Contents

Chapter 1

Introduction

Plant tissue culture technology has a great impact in the multiplication and conservation of medicinal and threatened plants and also production of plants with agronomically important traits for the crop improvement. This approach not only reduces the duration for micropropagation and selection of desirable traits but also facilitates raising of a large number of plants within a limited time and space.

It has wide applications in agriculture, horticulture, forestry and plant breeding etc. Plant tissue culture was exploited both for basic and applied aspects of plant research encompassing somatic embryogenesis, somaclonal variation, haploidy, mutagenesis, secondary metabolites, selection of cell lines of antibiotics and metabolic analogue resistance, drought resistance, somatic hybridization, genetic manipulation and molecular biology (Dix and Street, 1975; Lindsey and Yeomann, 1983; Mantell *et al.*, 1985).

Plant cell and tissue culture has emerged as a major tool in the study of an increasing number of applied and fundamental problems in the plant science and now it has become an important integral constituent of plant biotechnology. We have now reached a stage where we can manipulate cell and tissues more precisely and in a target based manner. The process of cell and tissue culture plays a major role in producing pathogen-free plants, synthetic seeds, secondary metabolites, somatic hybrids, cybrids and also provides basic requirement for genetic transformation to produce transgenics such as disease, insect, pest and herbicide resistance and stress tolerant plants.

Plant tissue culture is a technique for growing isolated cells, tissues and organs in a defined synthetic nutrient medium under aseptic conditions over prolonged periods in controlled environment.

The concept of cellular totipotency *i.e.*, ability to regenerate an entire plantlet from a single cell proposed by Haberlandt (1902) a pioneering German Botanist working on mechanically isolated mesophyll cells, emerged as corollary of the cell theory proposed by Schleiden (1838) and Schwann (1839). According to this concept "the genetic information necessary for the development of the entire organism is contained in all living cells, consequently when the cells are released from the developmental control, they are able to divide and differentiate into specific organs and whole organisms". Plant scientists took advantage of this concept and developed different types of media and methods over the years for *in vitro* culturing of explants which are induced to dedifferentiate in the proliferating mass of cells called callus.

Hanning (1904) initiated a new line of investigation involving the culture of embryogenic tissues, which later became an important applied area of investigation, using *in vitro* techniques. More promising results were obtained in the year 1908 by Simon, in the regeneration of bulky callus, buds, and roots from popular stem segments.

The callus production was the first major break through in plant tissue culture, achieved by White (1939), Nobecourt (1939) and Gautheret (1939) working independently with the explants of cambial tissue isolated from tobacco, carrot and carrot root respectively. Organogenesis from the callus can be induced by manipulating media composition and phytohormones from which whole plantlet is developed. The complete control of this process by the concentration of phytohormones represent a major area of interest in plant tissue culture.

From the time Haberlandt presented his paper in 1902 until about 1934 hardly any progress was made in the field of plant tissue culture. Plant scientists took advantage of the totipotency concept and developed media and method for tissue culture. The nutrient media used for most of the cultures were the formulations, developed in the early 1930's by White. Initially White used as medium containing inorganic salts, yeast extract and sucrose but later yeast extract was replaced by three B-vitamins *e.g.*, pyridoxin, thiamine and nicotinic acid (White, 1939), which was proved to be one of the basic media for a variety of cell and tissue culture studies.

The first complete chemically defined nutrient medium was developed by Murashige and Skoog (1962). The other media widely used for plant tissue culture studies were Lansmier and Skoog (1965) (LS), Gamborg's (B5), (Gamborg *et al.*, 1968) and SH media (Schank and Hildebrandt,1972). A standard culture medium consists of a balanced mixture of macro–and micro-nutrients, vitamins and plant growth regulators which play an important role in cell metabolism and cell membrane synthesis (Cocking, 1978).

In 1957, Skoog and Miller showed that the differentiation of roots and shoots in tobacco pith tissue cultures was a function of the auxin and cytokinin ratio and the organ differentiation could be regulated by changing the relative concentration of these two substances in the medium. But after the discovery of gibberellin and abscisic acid, altered this concept and so many factors are also responsible for control of organ formation. The specificity for organ type depends on the hormonal level which is governed by the exogenous supply and endogenous levels of phytohormones such as both auxins and cytokinins. In addition to these, environmental factors (notably light, temperature, nitrogen source, content of medium and other nutrient factors) and also genetic factors including physiology of donor plant and the age and type of the explant may influence.

Plant hormones/growth regulators have a special place in plant development. They are a structurally diverse group of low molecular weight molecules that have profound effect on plant growth when they are applied exogenously. Plant scientists took this advantage over identification of several hormones and media and they cultured several explants which are induced to differentiate into callus, callus and shoots/shoots. This callus may be differentiated into whole plants by modulating the concentration and combination of auxins and cytokinins. Among the growth regulators, auxins were proved to be essential for culturing plant tissue due to their effects on nucleic acid and protein metabolism (Gautheret, 1959; Rao and Swamy, 1972). Whereas cytokinins influence cell growth by promoting nucleic acid metabolism and synthesis of specific proteins required for cell division (Letham, 1968). A cytokinin bind to a specific receptor site in the cell to bring about organogenesis.

Morphogenesis can be obtained either directly from cultured explants or indirectly from callus or suspension cultures (George and Sherrington, 1984). In most cases, plant regeneration takes place by the formation of supplementary adventitious buds from the explants or *de novo* organisation of shoot and root meristem in the callus (Vasil *et al.*, 1982; Vasil, 1984). Callus can be increased substantially by subculturing periodically. The potential number of plants regenerated can thus be of very high from callus cultures. Several factors are responsible for plant regeneration (Murashige, 1974; Flick *et al.*, 1983). Later plant regeneration through adventitious shoot and root formation has been studied in a number of plants (Flick *et al.*, 1983; Huettman and Preece, 1993; Dhawan, 1993; Parthasarathy *et al.*, 2001; Collonnier *et al.*, 2001). Adventitious shoot regeneration *in vitro* provides a much higher rate of shoot multiplication. Adventitious shoot proliferation is the most frequently used multiplication technique in micropropagation system.

Another interesting area of research in tissue culture is clonal propagation through which plants are multiplied using axillary buds/nodal segments and shoot meristems as an explants. For rapid clonal propagation of plants in *in vitro* method, normally dormant axillary buds are induced to grow into multiple shoots by a judicious use of growth regulators cytokinins and or auxin + cytokinin combinations. Shoot number increases logarithemically with each subculture to give greatly enhanced multiplication rates. As this method involves only organized meristems, the plants obtained show no variability and allows recovery of genetically stable and true-to-type of progeny (Murashige, 1974; Hu and Wang, 1983). A large number of plant species have been successfully propagated by this method includes several crop plants, ornamentals, fruit crops, woody trees and medicinal plant species (Hu and Wang, 1983; Pierik, 1987; Mascarenhas and Muralidharan, 1989; Dhawan, 1993; Huettmann and Preece, 1993; Kathiravan and Ignacimuthu, 1999; Anwar and Siddiqui, 2000; Tiwari *et al.*, 2000; Sharon and D'Souza, 2000).

Meristem culture has been identified as an excellent technique for rapid multiplication and producing virus-free/pathogen-free plants. This meristem culture has been exploited successfully in herbaceous and medicinal plants (Bhojwani 1980; Wang and Hu, 1982; Mehra-palta, 1982; Bhaskaran *et al,* 1992; Nasir *et al,* 1997; Geetha and Shetty, 2000).

Clonal propagation aims at producing exact replicates (true-to-type) of the original plant selected for its desirable characters. Considerable progress has been made in recent years in clonal propagation of many plant species (Bhojwani, 1980; Wang and Hu, 1982; Yadav *et al.,* 1990; Malathy and Pai, 1998; Bais *et al.,* 2000).

Thus, micropropagation has several advantages over traditional methods of propagation for instance higher multiplication rates, lower requirements of space, propagation throughout the year and greater degree of control over chemical and physical environmental factors. Micropropagation provides a means of germplasm storage for maintenance of disease-free stock (Wilkins and Dodds, 1983; Withers, 1989). Fairly a large number of plant species has been propagated by *in vitro* techniques (Murashige, 1974; Narayana Swamy, 1977; Vasil and Vasil, 1980; Evans *et al.,* 1983; Parthasarathy *et al.,* 2001) but the degree of regeneration capacity varies considerably from species to species (Tisserat, 1987).

Since all the cells of the initial explant have identical genetic information, it was assumed that the plants regenerated from tissue culture would be genetically identical to one another and also to the donor plant, and would thus constitute a clone or a population of identical individuals. However, contrary to this expectation in the seventies and eighties of this century, it became an increasingly evident that regenerated plants were not always true-to-type (Evans and Sharp, 1983; Evans, 1989). One problem of clonal multiplication particularly of callus cultures is their tendency to genetic and chromosomal instability leading to somaclonal variation that may be transient or hertitable (monogenic or polygenic) which may improve the crops quantitatively and qualitatively. Therefore, these somaclonal variations could provide wide scope for producing new strain of crop plants with desired and useful traits (Evans and Sharp, 1986). Early reports on such culture-induced variability were dismissed as artifacts of the technique and no significance was attached to them. But Larkin and Scrowcroft (1981) on the basis of an extensive survey of the available literature on plants regenerated from tissue cultures, came to a conclusion that changes could be induced in the genetic material during *in vitro* culture leading to somaclonal variation. Interestingly, they found that such variations were noticed in almost all crop species tested, and often involved economically important traits such as male sterility, resistance to drought, abiotic stress and number of diseases. Thus, somaclonal variation has been exploited in sugar cane for increased sucrose content and resistance to eye spot disease (Heinz *et al.*, 1977; Krishna Murthy, 1982), in potato for growth habit, tuber colour and late blight resistance (Shepherd *et al.* 1980; Thomas *et al.*, 1982) and in tomato for resistance to *Fusarium* wilt (Evans, 1989). Later Nazeem *et al.* (1997) have developed the somaclones for tolerance to *Phytophthora* foot–rot resistance in black pepper.

Callus formation from explant tissue involves the development of progressively more random planes of cell division less frequent specialization of cells and loss of organized structure (Thorpe, 1980; Wagly, *et al.*, 1987). Cell suspensions are initiated by transfer of callus pieces into conical flasks with liquid medium, which are then placed on a gyratory-shaker to provide aeration to the cell (Rashid, 1988). As new cells are formed, they are dispersed into the liquid medium and become clusters and aggregates. Cells in suspensions can exhibit much higher rates of cell division than do cells in callus cultures. Thus cell suspensions offer advantages when rapid cell division or many cell generations are desired or when a more uniform treatment application is required such as during cell selection procedures. Muir (1953) succeeded in mechanically picking single cells from callus in suspension cultures. The success achieved thus for in single cell cloning created great enthusiasm for research on the prospects of raising whole plants from a single cell. An important break through was achieved in 1965 when Vasil and Hildebrandt observed that colonies arising from cloning of isolated cells of the hybrid *Nicotiana glutinosa* and *Nicotiana tabacum* regenerated plantlets.

Plant regeneration through somatic embryogenesis has several advantages over other routes of *in vitro* plant production and appears that most promising area of research for large scale production and rapid plant propagation. The first report of somatic embryogenesis in carrot tissue cultures *in vitro* published by Steward *et al.* (1958) and Reinert (1958). Somatic embryogenesis can be induced directly from a variety of explants or obtained indirectly by manipulating non-embryogenic callus *in vitro*. The totipotency of cells finds best expression in the formation of somatic embryos from single cells and their growth and development to form a complete plantlet (Ammirato, 1983a, b, 1987; Attree and Fowke, 1993; Finar, 1994). Somatic embryos believed to orginate from a single cell therefore plants derived from these tend to be genetically identical. Hence, it has great implications in tissue culture technology. Firstly, because somatic embryos have pre-formed root and shoot meristems, thereby reducing several of the labour intensive steps involved in the subculture, separation and rooting of individual shoots. Secondly, if the single cell origin of somatic embryos is universal as an increasing number of recent reports confirm, then potentially all single cells in cell suspensions could be induced to form embryoids in a prodigious way (Steward *et al.*, 1958, 1964; Jacobsen and Kysely, 1984). Cell suspensions are amenable to mechanization and automation, albeit with modification. This could have a major impact on methods of plant production. Already experimental system exists for encapsulation of somatic embryos with various hydrogels, thus producing an artificial seeds (Redenbaugh *et al.*, 1991; Attree and Fowke, 1993) which play an important role in germ plasm conservation of threatened and endangered medicinal and commercially important species. Thus, it is being viewed as an alternative method for mass propagation of several species in a short time.

Somatic embryogenesis has now been observed in a wide range of plants including those of the major crops (Ammirato, 1983a, b; Vasil and Vasil, 1984; Tiwari *et al.*, 1999; Mohan *et al.*, 2000; Hussain *et al.*, 2000). The literature shows that many of references describing the specific manipulations required to effect somatic embryo development from a variety of agronomically, medicinally and horticulturally important plants (Zimmerman, 1993; Roberts *et al.*, 1995; Mohan *et al.*, 2000; Chand and Singh, 2001). Efficient regeneration of plantlets from embryogenic callus is important step towards the genetic manipulation (Lorze *et al.*, 1988; Parrott *et al.*, 1991; McGranahan *et al.*, 1989). In addition to the development of somatic embryos from sporophytic cells, embryos have also been induced from generative cells such as in the classic work of Guha and Maheshwari (1964, 1966, 1967) with *Datura innoxia* microspores and Nitsch and Nitsch (1969) with *Nicotiana tabacum* micro spores.

Triploid embryos have also been observed in endosperm cultures of *Santalum album* (Lakshmisita *et al.*, 1980).

During conventional breeding programmes involving intraspecific, interspecific and intergeneric crosses, plant breeders are unable to produce viable hybrids due to the presence of barriers at various stages such as pollination, fertilization and embryo development. Finally it leads to the abortion of embryo. At this level, embryo culture plays a great role in plant breeding experiments for production of hybrid plants.

Embryo culture is the sterile isolation and growth of an immature or mature embryo *in vitro*, with the goal of obtaining a viable plant. The first attempt to grow the embryos *in vitro* was made by Hanning (1904) who obtained the viable plants from *in vitro* isolated embryos of two Crucifers *Cochleria* and *Raphanus* (Hanning, 1904). Embryos excised from developing seed at or near the mature stage are completely autotrophic. They germinate and grow on a simple medium with supplemented energy source. Very early in the history of tissue culture Laibach (1925, 1929) demonstrated the practical applications of zygotic embryo culture in the field of plant breeding, and he was the first to initiate work along these lines. Later Van Overbeek *et al* (1941) have demonstrated the stimulatory effect of coconut milk on embryo culture in *Datura*, which proved a turning point in the field of embryo culture. These studies gave impetus to further work in the area and to-date several hybrids have been raised through embryo cultures (Biggs *et al.*, 1986; Raghavan, 1977, 1986, 1994).

Anther culture technique plays a major role in the production of complete isogenic homozygous double haploids, compared to the conventional breeding technique. Guha and Maheshwari (1964) were first to produce plants in *Datura innoxia* through culture of anthers containing immature pollen, since then haploid plant production has been reported in many species (Dunwell, 1986). Later, the increasing importance of androgenesis in various plant species was emphasized by a number of workers (Bajaj, 1983; Morrison and Evans, 1988). The genetic variation produced after the anther culture is useful in plant breeding. Hence, it is possible to use gametoclonal variation for crop improvement (Evans *et al.*, 1984; Evans and Sharp, 1986; Hu and Zeng, 1986).

Microspore culture has a number of practical uses (Dunwell, 1985; Keller *et al.*, 1987; Knapp, 1991). It can reduce the time in obtaining new varieties and haploid embryos or doubled haploid plants. These androgenic haploid plants can also be used in mutation, biochemical, physiological and genetic engineering studies. Transgenic *Brassica napus* plants have been produced from pollen embryos with the use of *Agrobacterium tumefaciens* (Swanson and Erickson, 1989) and also by micro-injection of DNA into pollen embryos (Neuhaus *et al.*, 1987). Herbicide resistant plants have also been developed through *in vitro* selection of microspores (Baversodorf and Koit, 1987; Swanson *et al.*, 1988, 1989). Microspores derived embryos have been used in studies of lipid biosynthesis and storage (Taylor *et al.*, 1993). Haploids can be used to detect linkage and gene interaction as well as to estimate genetic variance and number of genes for quantitative characters.

The double haploid plants resulting from anther cultures are homozygous and breed-true. Production of doubled haploids from androgenic haploids is useful in plant breeding with improved efficiency for obtaining superior genotypes (Knapp 1991; Bjornastad *et al.*, 1993).

Plant cell cultures have been shown to produce a number of valuable secondary metabolites. These secondary metabolities are alkaloids, glycosides (steroids and phenolics), terpenoids and a variety of flavours, fragrances, perfumes, agrochemicals, commercial insecticides, shikonin and naphthaquinone (used as dye and pharmaceutical) which could be obtained from cell cultures (Fugita *et al.*, 1981; Yamamoto *et al.*, 1982; Green and Hedin, 1986; Waller, 1987; Flores and Medina-Bolivar, 1995).

Mutagenesis *in vitro* is another promising field for crop improvement (King, 1984). Explant mutagenesis *in vitro* may help to induce genetic variability and has been playing an increasingly important part in manipulating plant processes for crop improvement programme (Noyak and Mickel, 1987). Mutants may be of direct importance in plant breeding. Plant cells which do not respond to exogenous hormonal levels may be tried for regeneration using this technique. The effects of ionising rediation and chemical mutagens have been studied in many species *in vitro*. The literature on mutant selection has been reviewed by many workers (Chaleff, 1981, 1983; Maliga, 1980, 1984; Meins, 1983; Negruitiy *et al.*, 1984).

Attempts were made to utilise cell culture technology for developing genotypes which are resistant to salinity, high temperature, drought and various diseases and with improved amino acid spectrum (Widholm, 1974; Gengenbach *et al.*, 1977; Croughan *et al.*, 1981; Maliga, 1984; Chawla and Wenzel, 1987a, b; McCoy, 1987; Tal, 1994; Venkatachalam *et al.*, 1998).

Another interesting area of research in plant tissue culture is protoplast isolation, culture and fusion. For the first time Cocking (1960) isolated the plant protoplasts using enzyme cellulase. The first report of plant regeneration from isolated protoplasts was in *Nicotiana tabacum* (Takaba *et al.*, 1971). Since then, the plant regeneration from isolated protoplasts has been reported in many species (Evans and Bravo, 1983; Roest and Gilisen, 1989). Inorder to overcome sexual barriers, advances in protoplast fusion technology have made somatic hybridization techniques to facilitate the introduction of agronomcially important traits from wild relatives into cultivated species. This appraoch has an immense potential in transferring desirable agronomic traits using chemical or electrofusion for crop improvement (Razdan and Cocking, 1981; Jones, 1988; Parimerd *et al.*, 1988; Kobayashi *et al.*, 1996).

In most of the higher plants organelles are inherited uniparentally. Thus, combining cytoplasmic organelles, with different genetic traits is not possible by sexual hybridization in higher plants. Transfer of cytoplasmic organelles between plant species provide opportunities to investigate nuclear-organelle and organelle-organelle interactions. An advantage for the analysis of organelle transfer and interaction in cybrids is the presence of selectable and easily screened genetic markers in organelles (Gleba and Shlumukov, 1990; Medgyesy, 1990).

Mitochondria and plastids represent individual units of genetic information responsible for a number of vital agronomic and biochemical properties. The chloroplast DNA (Cp DNA) mystery and mitochondrial DNA (Mt DNA) mystery (Bendich, 1982) have been gaining much importance during the eighties, due to the exponentially growing information on the molecular organization of the genomes of plant organelles and also their role in crop improvement. Physical mapping of mitochondiral DNA and full sequencing of chloroplast DNA in several species mark this development. Results accumulating in the area of molecular biology of the Cp and Mt DNA offer an essential background to genetic analysis and manipulation of organelles with high efficiency.

The development of numerous plant cell and tissue culture techniques has opened new research avenues in crop improvement (Bottino, 1975; Carlson and Polacco, 1975). Plant tissue culture has been used as a means to modify the genetic make-up of crop plants in unconventional ways and thereby creating superior and novel genotypes. The potential to create unconventional genetic change has already been achieved in tobacco (Caplan *et al.*, 1985), tomato (Mc Cormick *et al.*, 1986), potato (Ooms *et al.*, 1987) and petunia (Meyer *et al.*, 1988).

In the production of transgenics also, tissue culture plays a vital role immediately after the transfer of desired gene/genes into a protoplast, cell, tissue or organ by using *Agrobacterium* mediated gene transfer or direct gene transfer method (particle bombardment, micro-injection, electroporation, lipofection, etc). Thus, it could complement conventional breeding techniques and the development of improved varieties in modern biotechnology. A lot of research has been going on for developing agronomically important transgenic plants using transformation technology.

Insect resistance was reported in tobacco (Vaek *et al*, 1987) and tomato (Fischhoff *et al*, 1987) first time using Bt genes. Disease resistance plants were developed in tobacco and *Brassica napus* introducing chitinase gene (Jones *et al*, 1988; Broglie *et al*, 1991). Viral resistant transgenic potato and tomato plants were developed using different types of coat protein genes. Reed *et al* (1995) developed and characterised the transgenic tomato plants that were delayed in fruit ripening. Herbicide tolerant transgenic plants have been reported in various food crops, vegetables, horticultural and ornamental species, but in cotton, flax, canola, corn and soybean these have already been released for commercial cultivation (James, 1997).

Transgenic plants conferring resistance against various abiotic stresses (teperature, water and salt) have been produced (Lee *et al*, 1995, 1999; Kishor *et al.*, 1995; Holmstrom *et al.*, 1996; Shen *et al.*, 1997; Hayashi *et al.*, 1997; Alia *et al.*, 1998; Murukami *et al*, 2000).

Very recent approach of modern biotechnology is the production of edible vaccines. The idea behind the edible vaccines is to have people take their dose by eating, as part of their diet, the plant (Pharma) that produces the vaccine. Many laboratories especially European and Western countries have been doing research for developing the plants (Pharma) which produce the edible vaccines to cure many diseases *viz.*, diabetes, hepatitis-B, diarrhea, cholera toxin-B, tooth decay etc. Therefore, transgenic plants show promise for use as low cost vaccine production systems. Attempts are being made to express many proteins of immunotherapeutic use at high levels in plants and to use them as bioreactors of the modern era (Sharma *et al*, 1999).

Thus it is obvious that the essential requirements for the successful application of plant propagation technology to agriculture is the capacity to regenerate elite plantlets. During the past decade, the demand for these plants has witnessed a steep rise. To meet the ever growing commercial requirements, the realization of *in vitro* multiplication of a large number of clonal plants with the improved characters has been gaining significance (Bais *et al.*, 2000). Thus, *in vitro* propagation can yield a large number of clonal plants for continuous plant establishment. It can also be important for germplasm conservation. Therefore in this book, an attempt has been made to review the present status of biotechnological research for improvement of *Solanum surattense*, an important in ayurvedic medicine.

Chapter 2

Introduction to the Species

The genus *Solanum* contains approximately 1000-1400 described species associated more than 3000 names. The genus is economically very important as several species are source of food, fodder and drugs. The fruits of brinjal are consumed as a vegetable all over India. Leaves of Indian shade (*S. nigrum*) have long been used in medicine for treatment of *Scrofulous* and *Dysuria* (Baily, 1949; Gopalaswamienger, 1951; Schery, 1952). Fruits are also used as medicine.

Many species of *Solanum* display resistance to fusarium wilt, verticellium wilt or bacterial wilt (Mochizuki and Yamakawa, 1979; Sakata *et al.* 1989).

The genus *Solanum* is medicinally important member of family Solanaceae. Several species of *Solanum* produce glyco-alkaloids which on hydrolysis and removal of sugar residues yield steroidal alkaloids solanine, solamargine and solasodine are widely distributed among the members of the genus. Various members of this genus have been exploited because of their high medicinal value. The steroidal glyco-alkaloids, *viz.*, solasodine and other closely related glycoalkaloids were found in the genus *Solanum* (Gu ilietti *et al.*, 1991). Solasodine serves as an important intermediate in synthesis of steroidal hormones (Butcher, 1977) and is a potential alternative to diosgenin as a precursor in the synthesis of steroidal hormones. Though solasodine is present in a number of *Solanum* species but berries of *S. avicuare, S.elaeagnifolium* and *S.khasianum* are particularly rich in it and are exploited commercially. The aglycones of these glyco-alkaloides are solanidine ($C_{27}H_{43}NO$) and solasodine ($C_{27}H_{43}NO$). These are isolated using the cell culture technology in *Solanum laciniatum* (Hosoda and Vatazawa, 1979; Chandler and Dodds, 1983). The *Solanum* glyco-alkaloids are toxic to animals when injected like the sapurine, they are surface active and haemolytic and possess antifungal and cytostatic properties (Manske and Halmes, 1970; Hoftmann, 1970). Chaturvedi *et al.* (1979) have determined the presence of solasodine from *Solanum khasianum*. Effect of boron and copper on growth and yield of solasodine content of *S. khasianum* was studied (Rosita, 1993). Solasodine based glyco-alkaloids solasonine and solamargin have altered the permeability of liposome membranes (Roddick *et al.*, 1992). Fewell *et al.*, (1994) have observed the inhibition of fungal growth by using solasonine and solamargin.

From the foregoing it is evident that many *Solanum* species are useful as vegetable and medicine. Hence, these species have been exploited in *vitro* thoroughly that the regeneration protocols for micropropagation, somatic embryogenesis, haploid production, somatic hybridization and genetic transformation have been established in some of the species. Chaturvedi and Sinha (1979) have developed the protocol for mass clonal propagation of *Solanum khasianum in vitro.* According to Ammirato (1983b) and Flick *et al.*, (1983) plant regeneration from *Solanum* tissue cultures may be achieved via organogenic or embryogenic pathways. Tejavathi and Bhuvana (1998) have studied the *in vitro* morphogenesis using root, hypocotyls, stem and leaf explants of *Solanum viarum.*

Plant regeneration through *in vitro* organogenesis (Sharma and Rajam, 1995; Fari *et al.*, 1995a, b; Magioli *et al.*, 1998) and somatic embryogenesis (Gleddie *et al.*, 1983; Kalloo, 1993; Yadav and Rajam, 1997, 1998) have been established in egg plant. *In vitro* shoot regeneration was also reported in other *Solanum* spp. *viz.*, *S. sisymbrifolium, S. xanthocarpum, S. aviculare, S. aethiopicum,* group *gilo, S. khasianum, S. indicum, S. tovum* and *S. nigrum* (Fassuliotis, 1975; Bhatt *et al.*, 1979; Kowzyk *et al.*, 1983; Gleddie *et al.*, 1985; Kashyap *et al.*, 1999). Resistant cell lines to pathogens, abiotic stress conditions and antibiotics have also been developed by using cell cultures in egg plant (Mitra *et al.*, 1981; Mitra and Gupta, 1989; Sadanandam and Farooqui, 1991; Asao *et al.*, 1992; Rao *et al.*, 1993, 1997; Ashfaq Farooqui *et al.*, 1997). Somatic hybridization has been successful to produce somatic hybrids between *Solanum melongena* and its wild relatives with the exception of *S. surattense* by using different types of fusion methods such as chemical and electrofusion to transfer nematode resistance, spider mites resistance, resistance to *Verticillium* wilt and shoot and fruit borer etc (Sihachakr *et al.*, 1988; Rotino *et al.*, 1995; Samoylov and Sink, 1996; Jarl *et al.*, 1999). The plant regeneration was also achieved through protoplast fusion between *S. malongena* and other related and wild species *viz.*, *S. sisymbrifolium, S. torvum, S. khasianum, S. gilo, S. nigrum* and *S. aethiopicum* (Kashyap *et al.*, 2003).

Solanum surattense Burm. f.

The species *Solanum surattense* Burm f. syn. *S. xanthocarpum* Schrad and Wendl ($2x = 24$) (Eng: Indian Solanum, Hindi : Baigan Kateli, Sanskrit : Kantakari, Malayalam : Kantakari Volutina), a common vegetable, is also used in herbal and ayurvedic formulations (Nayar *et al.*, 1989; Sivarajan and Balachandran, 1999).

It is a prostrate and diffuse herb, beset with yellow prickles allover, flowers bluish purple, fruit a glabrous, globular berry variegated with green and white stripes when young, yellow when mature and medicinally important (Gamble, 1986; Sivarajan and Balachandran, 1999) (Plate 1a&b). Fruits yield carpesteral, solasodine, solasonine and solanocarpine. They possess diuretic and antibacterial properties. The drug prepared from this plant is useful in bronchitis, muscular pain, enlargement of liver, spleen, blood cancer, vomiting and controlling stone in bladders (Ravindra Sharma, 2003).

The plant is extensively used in the treatment of fever, cough, asthma and constipation (Kirtikar and Basu, 200). The juice of berries of *S. surattense* was reported to be useful in *sore throat, hicough* and dropsy (Nadkarni, 1954; Chopra *et al.*, 1958). A clinical trial showed, decoction of *S. surattense* promote conception of females (Battacharjee, 1998). The leaf dust with flour is used as a repellant for *Trilobium castaneum* larvae (Hussain, 1995).

Plate 1a&b: Showing Plant Habit and Fruiting of *Solanum surattense*

S. surattense is an important therapeutic agent for dislodging tenacious phlegm and is extensively used in the treatment of cough, bronchitis, cholera, dysuria, diabetes and asthma. It is also useful in cases of influenza, enteric fever and allied conditions. The drug is diuretic, expectorant, febrifuge, cardiotonic, laxative, stimulant and is also used against difficult urination, bladder stones, rheumatism, sore-throat, enlargement of liver and spleen, vomiting and skin diseases (Sivarajan and Balachandran, 1999). Fumigation with the seeds is a reputed cure for toothache (Nadkarni, 1954). Roots, fruits and occasionally the whole plant are used in medicine. The important formulations using the drug are "*kantakari ghrtham, Putikaranjasavam, Suranadi leham* etc.," (Sivarajan and Balachandran 1999). The efficacy of *Kantakari* in the treatment of bronchial asthma (Shwas and Kas) and non-specific cough has been tried and also studied the pharmacognosy of this species. *S. surattense* is also used as one of the ingredients in the preparation of ayurvedic massage oil "*Kama virya*". Traditional ayurvedic (Indian) medicinal herb Kantakari used to treat sore throats, gum disorders, constipation and gas. All parts of this herb are used but must be used with knowledge and supervision by an herbal practitioner. **Bronchicyl** shows antispasmodic properties which has an expanding effect on the bronchi. *S. surattense* is used as one of ingredients of bronchicyl which stimulates the metabolism and do not just prevent the formation of mucous but also reduces inflammations. Fruits are also used as vegetable and reported to promote lowered blood cholesterol levels (Anonymous, 2006). The species is used in ayurvedic medicine as "*Dasamula*" for chest pains, cough, asthma, fever, etc., and fruit juice in ear ache (Nayar *et al.*, 1989).

Recently, it was also reported that the crude extracts of fruits and roots from the fresh plants of *S. surattense* showed larvicidal activity against vectors of malaria and dengue fever (Singh and Bansal, 2003). Although the species *S. surattense* has importance in ayurvedic medicine, a little work has been done *in vitro* (Prasad and Chaturvedi, 1978; Prasad *et al.*, 1998).

S. surattense was indiscriminately harvested from the natural flora for high value drugs which led to over exploitation that inevitably will result in virtual extinction of this species.

In view of over exploitation for its medicinal value and used as vegetable, Rama Swamy *et al.*, (2004, 2005a, b) have reported the improvement of the species by using biotechnological tools.

Chapter3

Callus Induction

The morphogenic response of a tissue in culture depends on the endogenous growth substances present in the culture medium. The callus formation from wounds is due to the effect of endogenous growth hormones was reported for the first time by Sinnot (1960). It is also dependent on the concentration of growth substances such as auxins alone or auxins-cytokinins present in the medium for inducing the callus. Callus is made of an amorphous aggregate of loose, parenchyma cells which proliferate from the mother cells (explant). When an explant is wounded, a callus is usually formed at the cut ends of that explant. Gautheret (1939) and Nobecourt (1939) for the first time reported the induction of callus, involving explants of cambial tissues isolated from carrot root respectively. The general growth characteristic of a callus involves a complete relationship between the plant material used to induce callus and the composition of the medium and the culture conditions (light, temperature etc) during the incubation period (Murashige, 1974). All the plant tissues containing living cells may be induced to form dedifferentiated mass of cells, callus, but tissues may vary in terms of their lag period before an active growth is induced. Mostly all the multicellular plants are potential source for initiation of callus (Yeoman and Macleod, 1977).

The application of plant tissue culture methods in crop improvement depends upon the induction of viable callus cultures and capable maintenance *in vitro* (Vasil *et al*, 1979). Establishment of callus cultures from the explant can be divided into three stages, induction of cell division, continued proliferation and structural and physiological dedifferentiation. Some calli are heavily liquified and hard in texture where as others break easily into small fragments and readily separated are termed as "friable callus". Generally in tissue cultures two types of calli are recognised. They are : (i) embryogenic callus which is smooth, compact and cream or yellowish in colour and consists of isodiametric cells, (ii) non-embryogenic callus that is generally rough, crystalline, yellow to brown in colour and consists of tubular cells. The nature of the callus tissue, its texture, compactness, friability and coloration also depends on the genotype, age of the explant including the plant and even the season. The callus growth within a plant species is also dependent on various factors such as the original position of the explant within the plant and the growth conditions referring physiological status of the plant.

Callus is somewhat an abnormal tissue which has the potentiality to produce normal roots or shoots and embryoids which develop into complete plantlets depending on the level of phytohormones *i.e.*, auxin and cytokinins present in the nutrient medium.

Induction of callus from an explant plays a great role in tissue culture technology for production and multiplication of somaclonal variants with agronomically important characters, isolation of pharmaceutically and therapeutically important chemicals (secondary metabolites) from medicinally important plants using cell suspension cultures and also for conservation of the species. In view of all these, an attempt has been made by Rama Swamy and coworkers (unpublished) to establish the efficient protocol using various growth regulators in inducing the callus from different explants in *S. surattense* a medicinal herb.

Callus induction ability of different explants such as cotyledon, hypocotyl and leaf was investigated by using varying concentrations of different auxins individually. Callus proliferation was initiated at the cut surfaces of the explants studied and later it covered the entire surface. Both color and texture of the callus also varied with growth regulators supplemented. The results are presented in Tables 3.1–3.3 and shown in Plate 2. The explants *viz.*, cotyledon, hypocotyl and leaf cultured on MS medium supplemented with different concentrations (1.0 mg/L to 5.0 mg/L) of auxins such as 2,4-D, IAA and NAA individually exhibited initiation of callus after 10days of incubation while it took 12-15 days in cotyledon cultures.

TABLE 3.1
Effect of 2, 4-D on Callusing Ability of Different Explants of *S. surattense*

Sl.No.	Explant	2,4-D Concentration (mg/L)	% of Cultures Responding	Morphology	Callusing Response
1.	Cotyledon	1.0	100	White-compact	++
		2.0	100	White-compact	+++
		3.0	100	White-compact	+++
		4.0	100	White-compact	++
		5.0	100	White-compact	+
2.	Hypocotyl	1.0	80	White-compact	+++
		2.0	65	White-compact	+++
		3.0	60	Cream-friable	+++
		4.0	55	Brown-compact	++
		5.0	50	Grey-compact	+
3.	Leaf	1.0	90	White-compact	+++
		2.0	85	White-compact	+++
		3.0	80	Cream-friable	+++
		4.0	70	Brown-compact	++
		5.0	60	Brown-compact	+

Relative amount of callus formation: —: No; +: Low; ++: Moderate; +++: High

On 2,4-D supplemented medium, early induction of callus was observed except at 1.0 mg/L in all the explants tested. Callus induction was observed from all the explants cultured and in all the concentrations of 2,4-D. 100 per cent callusing response was recorded in cotyledons, cultured in all the concentrations of 2,4-D. Maximum percentage of response was observed in cotyledon cultures followed by leaf cultures and hypocotyl at 1.0 mg/L 2,4-D (Table 3.1). As the concentration of 2,4-D was increased, there was a gradual decrease in the response in hypocotyl and cotyledon cultures but it was interesting to note that 100 per cent response was observed in all the concentrations of 2,4-D in cotyledon explants. High amount of callus was induced on 2,4-D in all the explants except at 4.0 and 5.0 mg/L 2,4-D. Morphology of callus was also found to be varied at different levels of 2,4-D (Table 3.1). Friable callus was induced at 3.0 mg/l 2,4-D in leaf and hypocotyl cultures whereas compact calli were induced in almost all the concentrations of 2,4-D.

TABLE 3.2
Effect of IAA on Callusing Ability of Different Explants of *S. surattense*

Sl.No.	Explant	IAA Concentration (mg/L)	% of Cultures Responding	Morphology	Callusing Response
1.	Cotyledon	1.0	68	White-compact	+
		2.0	60	White-friable	++
		3.0	58	Cream-friable	++
		4.0	50	Brown-friable	+
		5.0	—	—	—
2.	Hypocotyl	1.0	50	White-compact	++
		2.0	44	White-friable	++
		3.0	36	Cream-friable	+++
		4.0	30	Cream-friable	+
		5.0	16	Grey-compact	+
3.	Leaf	1.0	70	White-compact	++
		2.0	68	Cream-embryogenic	++
		3.0	60	Cream-embryogenic	+
		4.0	30	Brown-compact	+
		5.0	—	—	—

Relative amount of callus formation: —: No; +: Low; ++: Moderate; +++: High

Callusing efficiency of cotyledon, hypocotyl and leaf explants cultured on MS medium containing various concentrations of IAA is presented in Table 3.2. High percentage of response was observed at low level of auxins used in all the explants but at 5.0 mg/L IAA, the callus induction was inhibited in cotyledon and leaf explants. Whereas less percentage of response was observed at 5.0 mg/L IAA in hypocotyl cultures. High amount of callus was induced at 3.0 mg/L IAA in hypocotyl cultures compared to all other concentrations of auxin and explants too. Embryogenic callus was found at 2.0 and 3.0 mg/L IAA in leaf cultures. Friable callus was observed at all other concentrations of IAA in cotyledon, hypocotyl and leaf cultures (Plate 2a) except at 1.0 mg/L in cotyledon and at 4.0 mg/L in leaf explants.

TABLE 3.3
Effect of NAA on Callusing Ability of Different Explants of *S. surattense*

Sl.No.	Explant	NAA Concentration (mg/L)	% of Cultures Responding	Morphology	Callusing Response
1.	Cotyledon	1.0	90	White-compact	++
		2.0	92	White-nodular	++
		3.0	98	White-friable	++++
		4.0	89	Cream-friable	++++
		5.0	80	Cream-friable	+++
2.	Hypocotyl	1.0	70	White-compact	++
		2.0	69	Cream-compact	++
		3.0	68	Cream-friable	++++
		4.0	50	Cream-friable	++
		5.0	42	Brown-friable	++
3.	Leaf	1.0	90	White-nodular	+++
		2.0	95	Green-nodular	+++
		3.0	98	Brown-nodular	++++
		4.0	100	White-friable	++++
		5.0	85	White-friable	++

Relative amount of callus formation: —: No; +: Low; ++: Moderate; +++: High

A frequency of 70 per cent response was found at 1.0 mg/L IAA in leaf cultures. As the concentration of IAA was increased to 2–4 mg/L, the percentage of responding cultures decreased. Hypocotyl response was found to be less compared to cotyledon and leaf cultures on MS medium containing IAA but callusing efficiency was high at 3.0 mg/L IAA compared to all other concentrations and explants too.

Effect of NAA on callusing ability of cotyledon, hypocotyl and leaf explants is shown in Table 3.3. Highest percentage (100 per cent) of response was observed at 4.0 mg/L NAA in leaf cultures whereas less percentage of responding cultures were noted in all the concentrations of NAA used in hypocotyl explants. Cotyledon and leaf explants were responded well at all levels of NAA. Very high amount of callus was produced from cotyledon and leaf cultures on MS medium supplemented with 3.0 and 4.0 mg/L NAA. While at 3.0 mg/L NAA more amount of callus was induced in hypocotyl explants. The callus induction was moderate at high concentrations of NAA in leaf and hypocotyl explants. Induction of nodular callus was observed (Plate 2b) in both cotyledon and leaf explants on MS medium containing different concentrations of NAA except at low concentration of auxin in cotyledon explants and high concentration in leaf explants. Whereas in hypocotyl cultures, friable callus was produced on 3.0–5.0 mg/L NAA. Friable callus was also induced at 3.0–5.0 mg/L NAA from cotyledon explants and at 4.0 and 5.0 mg/L from leaf explants.

The callus was induced from all the explants *viz.*, hypocotyl, cotyledon and leaf in all the concentrations of auxins 2,4-D, IAA and NAA with the exception of 5.0 mg/L IAA in *S. surattense*. At this high concentration of IAA, the morphogenetic response was inhibited in cotyledon and leaf explants. Callusing efficiency was found to be higher in leaf explants followed by cotyledon and hypocotyl explants of *S. surattense*. This difference in callusing ability suggests the presence of different levels of endogenous hormones in the tissues. The auxins such as NAA, IAA and 2,4-D alone suppressed shoot bud formation in all the explants studied but promoted callusing.

Among the auxins tested, NAA induced the high yield of callus followed by IAA and 2,4-D. Similarly, Omar (1988) observed the same findings with NAA in *Rhazya stricta*, a medicinal plant. Whereas Skoog and Miller (1957) and Kitamura (1988) recorded the higher callogenesis with auxin, NAA in combination with either BAP or Kn. These auxins and cytokinins both together act synergistically to promote either cell division or expansion

Plate 2a: Induction of Friable Callus on MS + 3.0 mg/L IAA from Cotyledon Explant in *S. surattense.*

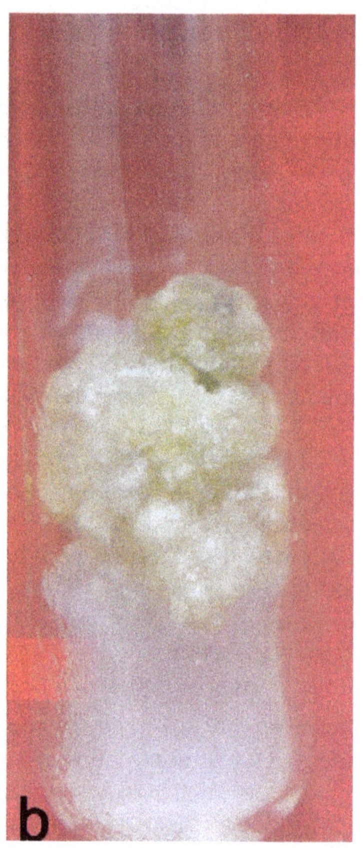

Plate 2b: Induction of Green Nodular Callus on MS + 2.0 mg/L NAA from Leaf Explant

depending upon other factors within the cell which react with these hormones (Setterfield, 1963; Gamborg *et al.*, 1977).

The auxin 2,4-D has been determined as a potent callus inducing auxin in studies with many plant species: *Capsicum* (Gunay and Rao, 1978; Philips and Hubstenberger, 1985), *Cucumis sativus* (Rajasekharan *et al.*, 1983). Whereas in *S. surattense* 2,4-D induced less amount of callus proliferation compared to all other auxins used in all the explants studied. Praveen *et al.* (2001) have studied the callusing ability of different explants in *Strychnos potatorum* a medicinal forest tree on various growth regulators *viz.*, IAA, NAA and 2,4-D. They observed that the maximum callusing efficiency on MS medium containing 2,4-D in contrast to *S. surattense.*

Tejavathi and Bhuvana (1998) have also observed the callusing ability of different explants in *Solanum viarum* using auxins NAA and 2,4-D. Among these auxins, NAA (2 mg/L) was found to be the best to induce callus from hypocotyl, while 2,4-D (3 mg/L) either alone or with CM (coconut milk) (10 per cent) elicited callus formation from root, stem and leaf explants (Tejavathi and Bhuvana, 1998). Kumari and Kumar (1995) have observed the friable callusing and rhizogenesis in the explants cultured on a medium containing IAA, NAA and 2,4-D at 1–25 mM range of concentration in *Thevetia peruviana.* Shahzad *et al.* (1999) have observed the callus induction on MS medium supplemented with NAA (2 mg/L) and 2,4-D (2 mg/L) in leaf cultures of *Solanum nigrum.* They found the faster proliferation and very high yield of callus on 2,4-D compared to NAA.

For induction of callus from any explant auxins are essential but depending upon the species callusing ability is enhanced by adding low level of cytokinins in the medium along with the auxins. Auxins and cytokinins act synergistically for promoting the callus induction at a particular level of concentrations. This phenomenon was found in various plant systems. Abundant callus was initiated in scented *Pelargonium* on 2,4-D in combination with Kn (Jacqueline and Charlwood, 1986). It was also interesting to note that 2,4-D in combination with Kn showed efficient than 2,4-D alone in inducing abundant callus in different species of *Capsicum* (Venkataiah, 2001). It was also observed in *Momordica dioica* the essentiality of both auxins and cytokinins for induction of callus (Hoque *et al.*, 2000). Similar findings were also found in *Strychnos potatorum* on MS medium amended with 2,4-D in combination with BAP/Kn (Praveen *et al.*, 2001).

Callus produced from different explants showed variability in texture, form and coloration. This difference is dependent upon the responses of plant tissues to various growth promoting substances. Thus, successful callus induction depends upon various factors such as composition of the nutrient medium, hormonal balance besides the type, age and genotype of the explant (Huang and Murashige, 1976; Narayana Swamy, 1977).

Thus it is evident that among the auxins, NAA was found to be potent for callus proliferation followed by IAA and 2,4-D in *S. surattense.* Proliferation in callus was faster and a very high yield of callus mass was achieved within 4 weeks on NAA followed by IAA and 2,4-D. Maximum amount of callus production was observed on MS medium supplemented with 3.0 and 4.0 mg/L NAA with highest percentage of responding cultures. Callusing ability was found to be more in leaf cultures followed by cotyledon and hypocotyl explants. So, among the explants tested leaf explants are highly efficient for inducing the callus production in *S. surattense*. The callus produced from 2,4-D was white and brown/gray compact whereas from IAA and NAA, the callus was friable/embryogenic in almost all the concentrations and explants used indicating the capability for regeneration.

Since Giulietti *et al*. (1991) have used the cell suspension culture technique for isolating *Solasodine* in *Solanum eleagnifolium* from calli induced on different auxins such as IAA, 2,4-D, NAA, IBA and 2,4,5-T. The same technology can also be used to isolate the different glycoalkaloids (Solasonine, Solamargine, Solasodine) from *in* vitro cultures of *S. surattence.*

Chapter 4

Plant Regeneration/ Organogenesis In vitro

Plant regeneration/organogenesis is an important aspect of the tissue culture. This phenomenon plays a great role *in vitro* plant micropropagation and conservation of the germ plasm.

Plant regeneration from tissue culture may be achieved via organogenesis or embryogenesis (Ammirato, 1983a,b; Flick *et al*, 1983). Organogenesis is the process by which a cell, or group of cells, differentiates to form organs. Organogenesis refers equally to the formation of roots or shoots, but since recovery of plants is usually the objective, the formation of shoots is of greater interest. Organogenesis is commonly induced by manipulation of exogenous phytohormone levels and occurs either directly from explant tissue or from callus. Often, what is referred as direct organogenesis is, infact, shoot proliferation or micropropagation from pre-existing meristems instead of *de novo* formation of a meristem. Even in callus mediated organogenesis/regeneration organ-forming capacity limited to "primary" callus indicates the potential existence of meristems embedded in the original explant. However, the degree of regeneration varies considerably from species to species (Tisserat, 1987).

In vitro organogenesis / regeneration can be obtained by 3 ways:

1. Through callus mediated regeneration
2. Somatic embryogenesis and
3. Direct organogenesis/shoot bud proliferation/ regeneration based on the auxin-cytokinin ratio in the nutrient medium.

Regenerative ability depends on the genotype of the donor tissue, source of explant and growth regulators such as auxins or auxin-cytokinin combination and also the environmental conditions during the incubation period. The influence of these factors in *in vitro* regeneration was reported in *Lycopersicon esculentum* (Tan *et al*, 1987), *Brassica* (Narasimhulu and Chopra, 1988; Khehra and Mathias, 1992), *Glycine* (Amberger *et al*., 1992), *Solanum melongena* (Sharma and Rajam, 1995) and *Dianthus caryophyllus* (Kallack *et al*., 1997) and *Capsicum annuum* (Venkataiah and Subhash, 2001).

A) CALLUS MEDIATED REGENERATION

A dedifferentiated tissue such as callus can redifferentiate a variety of organs under appropriate culture conditions. Organs arise *de novo* and not necessarily from pre-existing initials in a tissue explant. The callus mass consists predominantly of enlarged, vacuolated parenchyma cells. Subsequently, differentiation of the several specialized types of cells prevails, although less diverse than that of an intact plant (Gautheret, 1966). Growth regulator concentration in the culture medium is critical for morphogenesis. By varying the growth regulator levels and types, morphogenesis pattern can be changed. As mentioned earlier, medium containing high auxin levels induce the callus formation. Lowering the auxin and increasing the cytokinin concentration or presence of only cytokinin in the medium can generally induce shoot organogenesis / regeneration from callus. Shoot regeneration from callus cultures derived from different explants was reported in various species : *Piper longum* (Philip *et al*., 2000), *Momordica dioica* (Hoque *et al*., 2000). Successful plant regeneration from callus cultures in some *Solanum* species was also achieved (Gleddie *et al*., 1985; Baburaj and Gunasekaran, 1994; Shahzad *et al*., 1999).

During the present investigations freshly isolated callus showed greater morphogenetic response for induction of shoots. The morphogenetic callus continued to generate shoots upon subculture.

The number of shoots produced from callus cultures varied with the hormones added to MS basal medium in *S. surattense.* Auxin in combination with a cytokinin produced maximum number of shoots with a greater frequency when compared to cytokinin as a sole growth regulator (Figure 4.1).

Callus derived from leaf explants when cultured on cytokinins alone or in combination with auxins showed the shoot bud induction after one week of culture.

A frequency of 50 per cent cultures responded at 1 mg/L BAP with an average number of shoots 12.4 ± 0.3 per explant. At 3.0 mg/L BAP, 63.0 per cent cultures responded with a maximum number of shoots 14.9 ± 0.7 (Plate 3a). Higher levels of BAP *i.e.,* at 4.0 mg/L and 6.0 mg/L produced less number of shoots 11.7 and 11.2 with a lower frequency of 47.0 and 26.0 response respectively. Shoot bud suppression was observed at 8.0 mg/L BAP. Kinetin was found to be less responsive compared to BAP. A high frequency of 56.2 per cent cultures responded with a maximum number of shoots/explant (13.9 ± 0.5) at 3.0 mg/L Kn. Low and high levels of Kn showed less response as observed in the case of BAP. As the concentration of Kn was increased to 8.0 mg/L, shoot bud initiation was completely suppressed.

After the study of cytokinin as a sole growth regulator on shoot regeneration from callus cultures, auxin was taken in combination with cytokinin to study the effect on regeneration ability. The concentration of auxin IAA was kept constant at 0.5 mg/L and different concentrations of BAP (1, 2, 3, 4, 6 and 8.0 mg/L) were added to the MS medium. Lower level of BAP, 1.0 mg/L induced less number of shoots (16.5 ± 0.54) with frequency of 46.5 per cent cultures responding. Whereas at 3mg/L BAP, 60.5 per cent cultures responded and maximum number of shoots/explant (20.5 ± 0.53) were recorded (Plate 3b). As the concentration of BAP increased to 4.0 and 6.0 mg/L the percentage of responding cultures was reduced to 58.0 and 45.0 and there was also a decrease in the number of shoot bud proliferation 15.6 and 13.5 respectively. Only green nodular callus was produced without shoot bud initiation at 8 mg/L BAP.

The calli cultured on MS medium supplemented with 0.5 mg/L IAA in combination with various concentrations of Kn showed less percentage of responding cultures compared to the combination

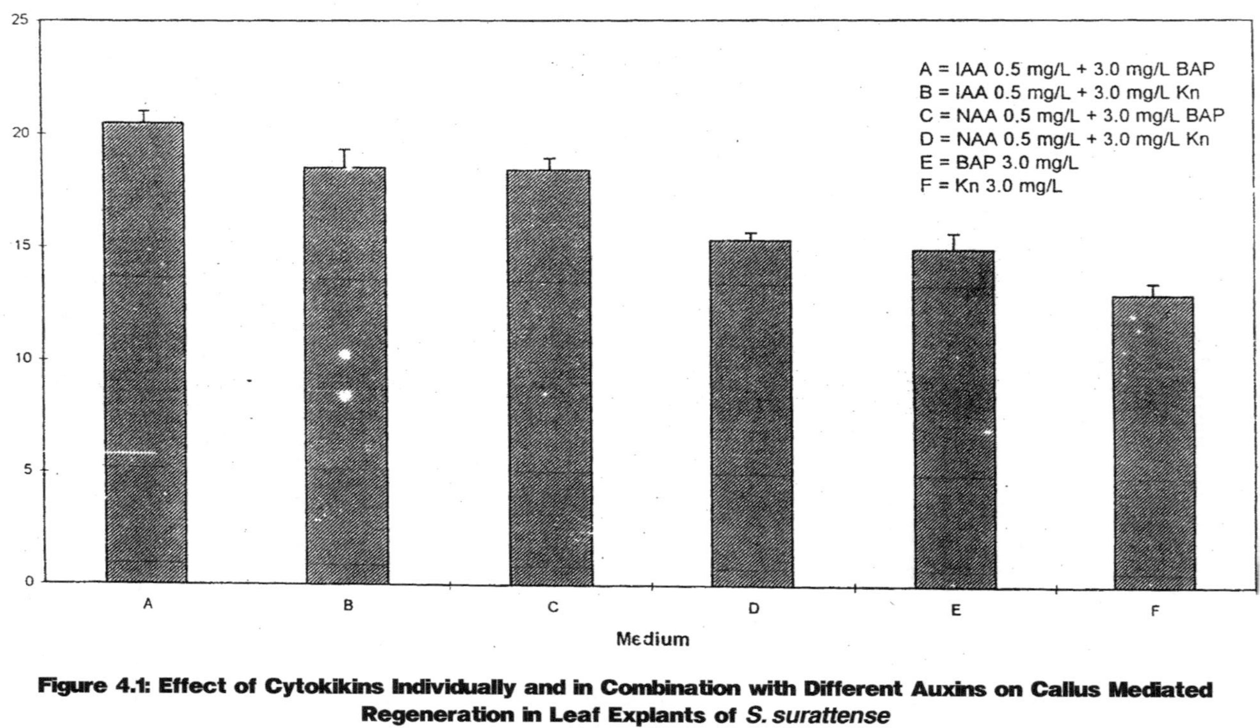

Figure 4.1: Effect of Cytokikins Individually and in Combination with Different Auxins on Callus Mediated Regeneration in Leaf Explants of *S. surattense*

Plate 3a: Callus Derived on MS + 4.0 mg/L NAA from Leaf Explant Cultured on 3.0 mg/L BAP (Note the organogenesis after 10 days) in *S. surattense.*

Plate 3b: Shoot Regeneration from Callus Derived from Leaf Explant on MS + 4.0 mg/L NAA, Cultured on MS + 0.5 mg/L IAA + 3.0 mg/L Kn after 6 weeks

Plate 3c: Direct Proliferation of Shoot Buds on MS + 1.0 mg/L IAA + 3.0 mg/L BAP from Leaf Explant

Plate 3d: Profusly Developing High Frequency Multiple Shoots from Leaf Explants on 0.5 mg/L NAA + 3.0 mg/L BAP (Note hundreds of shoots formation).

Plate 3e: Induction of Roots on □ MS + 0.5 mg/L IAA from Leaf Regenerant

of BAP. Different concentrations of Kn (1, 2, 3, 4, 6 and 8 mg/L) were added to MS medium in combination with 0.5 mg/L IAA. At 1 and 2 mg/L Kn 46 per cent and 60 per cent cultures responded with an average 14.5 and 15.6 shoots/explant respectively. A maximum of 18.5 shoots/explant was recorded at 3 mg/L Kn with a frequency of 85 per cent cultures responding. Number of shoots considerably were reduced as the concentration of Kn was increased. Responding cultures were also reduced after 3.0 mg/L Kn as observed on BAP. Complete suppression of shoot bud induction was found at high concentration of Kn.

Similarly, the callus derived from leaf explants was cultured on 0.5mg/L NAA in combination with different concentrations of BAP or Kn. Low level of BAP (1.0, 2.0 mg/L) induced less number of shoots/explant 12.3 and 13.3 with less percentage of responding cultures. At 3.0 mg/L BAP, 54 per cent cultures responded and a maximum number of shoots (18.4) per explant were recorded. It was also noted that as the concentration of BAP was increased to 4 and 6 mg/L the percentage of responding cultures reduced to 27 per cent and 23 per cent and also a decrease in number of shoots 13.2 and 12.0 respectively. Only green nodular callus was induced without any shoot bud initiation at 8 mg/L BAP. Regeneration of shoot buds from callus cultures on 0.5 mg/L NAA in combination with different concentrations of Kn was found to be less response compared to BAP. Low levels of Kn induced less number of shoots/explant showing less percentage of responding cultures. At 3.0 mg/L Kn maximum number of shoots was produced. Higher levels of Kn (4.0 and 6.0 mg/L) showed less response, producing less number of shoots/explant. At 8.0 mg/L only callus proliferation was observed without any shoot bud initiation as observed on BAP.

B) DIRECT REGENERATION

In vitro plant regeneration is an essential prerequisite for genetic engineering and the production of transgenic plants for crop improvement.

In direct regeneration, shoots/shoot buds, arise directly from an explant without an intervening callus phase. There have been numerous reports in recent years on regeneration of plants from various types of excised explants, *viz.*, cotyledon, hypocotyl, leaf, root, embryos etc. of many species of Solanaceae (Subhash and Christopher, 1988; Sharma and Rajam,1995; Fari *et al.*, 1995a).

Rama Swamy *et al* (pers. comm) have established the protocols for an efficient direct regeneration from different explants *viz.,* hypocotyl, cotyledon and leaf in *S. surattense.*

The role of cytokinins and auxin-cytokinin combinations on direct plant regeneration and adventitious bud induction from different explants *viz.*, hypocotyl, cotyledon and leaf was studied inorder to findout the efficient protocol and potential explant in *S. surattense* (Rama Swamy *et al*,2006a). All the explants were cultured on MS medium fortified with different concentrations of cytokinins alone and auxins (0.5/1.0 mg/L) in combination with various concentrations of cytokinins such as BAP/Kn (1.0–10.0 mg/L). These explants were enlarged 3-4 fold within one week of culture initiation. Morphogenic changes were apparent after 4 weeks of culture. The explants *viz.*, hypocotyl, cotyledon and leaf developed shoot primordia in large numbers directly from all cut surfaces in all the concentrations and combinations of growth regulators used except at 10.0 mg/L BAP/Kn and even in combination with 0.5/1.0 mg/L IAA. At this concentration, direct organogenesis was found to be completely inhibited. The results are presented in Figures 4.2-4.6.

Cotyledon explants

Effect of various concentrations of cytokinins such as BAP and Kn alone and in combination with various concentrations of auxin IAA was studied on direct multiple shoot bud induction in cotyledon explants of *S. surattense* (Figure 4.2). Direct adventitious shoot regeneration on MS medium containing various concentrations of BAP (1-8 mg/L) was observed with varied results. Highest responding cultures with maximum frequency of multiple shoot bud induction was observed at 3.0 mg/L BAP (29.0 shoots/explant) followed by 4.0 mg/L BAP (25.0 shoots/explant). At high concentration of BAP the percentage of responding cultures and shoot bud proliferation were reduced.

Direct multiple shoot bud induction from cotyledon explants cultured on various concentrations of Kn (1-8 mg/L) was observed. High percentage (89) of responding cultures was found at 4.0 mg/L Kn compared to all other concentrations tested. Whereas more number of shoots were regenerated from cotyledon explants at 3.0 mg/L Kn (25.7 ± 0.35 explant) followed by 4.0 mg/L Kn in which profuse rhizogenesis was also observed. At high concentration of

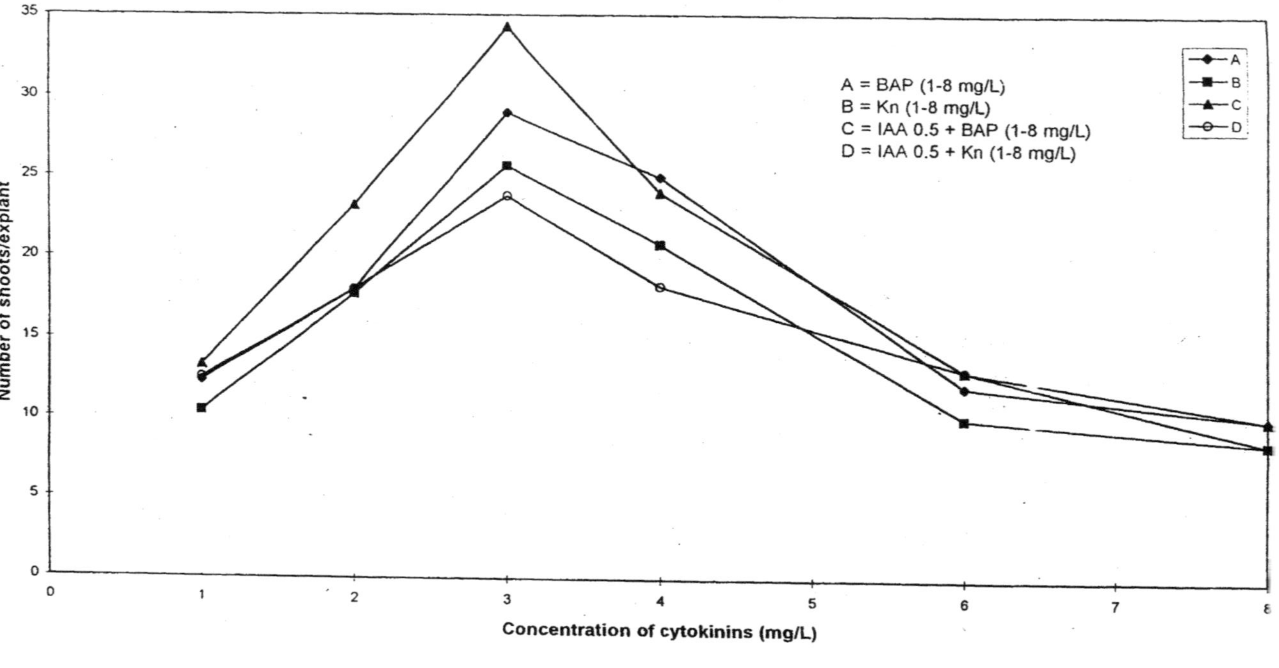

Figure 4.2: Effect of Increasing Concentration of BAP and Kn Individually and also IAA+BAP and IAA+Kn on Direct Shoot Bud Proliferation from Cotyledon Explants of *S. surattense*

Kn less frequency of shoot organogenesis was observed. On the whole, the response of BAP as a sole growth regulator was found to be maximum in inducing shoot organogenesis of all the concentrations tested in comparison to Kn (Figure 4.2).

Influence of auxin-cytokinin combination such as IAA (0.5/1.0 mg/L) + BAP (1-8 mg/L) and IAA (0.5/1.0 mg/L) + Kn in cotyledon explants showed variable response. Auxin, 0.5 mg/L IAA was taken in combination with cytokinin, BAP (1 to 8 mg/L) showed maximum percentage (95 per cent) of responding cultures and high frequency of shoot regeneration (34.3 shoots per explant) at 3.0 mg/L BAP. As the BAP concentration was increased to 8 mg/L, gradually the induction of shoot organogenesis was found to be reduced. The cotyledon explants when cultured on MS medium fortified with 1.0 mg/L IAA in combination with 1,2,4,6 and 8 mg/L BAP showed maximum shoot induction frequency at 3.0 and 4.0 mg/L BAP (14.0 and 13.5 shoots/explant). At 4.0 mg/L BAP in combination with 1.0 mg/L auxin IAA produced roots in addition to multiple shoots induction. The response in inducing the shoot organogenesis was found to be lesser after enhancing the concentration of auxin from0.5 to 1.0 mg/L.

Similarly, cotyledon explants were cultured on MS medium supplemented with IAA (0.5/1.0 mg/L) in combination with various concentrations of Kn (1-8 mg/L). The results showed the direct shoot regeneration in all the concentrations and combinations tested. Cotyledons cultured on 0.5 mg/L IAA in combination with different levels of Kn showed highest number of shoots (23.8 ± 0.35) at 3.0 mg/L Kn and more percentage (98) of responding cultures. But at high level of Kn, shoot bud induction was reduced along with the response too. When the concentration of IAA was increased to 1.0 mg/L, the induction frequency of shoot organogenesis was gradually decreased. From the results it is clear that BAP showed superiority over Kn when it was used individually or along with the two different concentrations of auxin for inducing direct multiple shoot regeneration (Figure 4.2).

Hypocotyl explants

Hypocotyl explants cultured on MS medium containing various concentrations of cytokinins BAP/Kn alone showed the direct shoot regeneration. Highest percentage (98) of responding cultures was observed at 3.0 mg/L BAP followed by 2.0 mg/L and gradually reduced as the level of BAP was increased. Maximum number of shoot organogenesis (21.0 ± 0.75) was also found at 3.0 mg/L BAP whereas the shoot bud induction was decreased at high level of BAP.

Similarly, hypocotyl explants were cultured on MS medium fortified with various levels of kinetin (1-8 mg/L). Maximum frequency of shoot bud induction (20.3 ± 0.25/explant) was noted at 3.0 mg/L Kn compared to all other concentrations tested. Shoot regeneration capacity was gradually decreased at high concentration of Kn as observed in BAP. The response of both the cotyokinins was found to be almost same in all the concentrations triggering the shoot buds proliferation in hypocotyl explants (Figure 4.3).

To findout the efficacy of auxin-cytokinin combination, the hypocotyl explants were cultured on MS medium augmented with IAA (0.5/1.0 mg/L) in combination with various concentrations of BAP/Kn (1-8 mg/L). Direct shoot bud proliferation was found in all the concentrations and combinations of growth regulators used.

Hypocotyl explants cultured on MS medium containing 0.5 mg/L IAA in combination with 1-8 mg/L BAP showed maximum responding cultures and more number of shoots/explant (13.5 ± 0.2) at 3.0 mg/L BAP. Average number of shoots production has been gradually decreased at high concentrations of BAP. When hypocotyl explants were cultured on 1.0 mg/L IAA in combination with various concentrations (1, 2, 3, 4, 6 and 8 mg/L) of BAP showed high number of shoot buds/explant (15.3 ± 0.35) compared to all other concentrations of BAP and Kn as well. But at high level of BAP the response was decreased. The results show that for high frequency bud induction in hypocotyl cultures the auxin level was increased to 1.0 mg/L.

Hypocotyl explants cultured on MS medium fortified with 0.5/1.0 mg/L IAA and different concentrations of Kn (1-8 mg/L). Direct adventitious shoot bud induction and plantlet formation was observed on 0.5 mg/L IAA in combination with 1-8 mg/L Kn. High frequency of shoot organogenesis with highest responding cultures was found at 3.0 mg/L Kn compared to all other concentrations of Kn. When the auxin concentration was increased to 1.0 mg/L IAA along with the different concentrations of Kn showed almost the same efficacy of Kn but at high concentration of Kn, the average number of shoot regeneration per explant was gradually decreased.

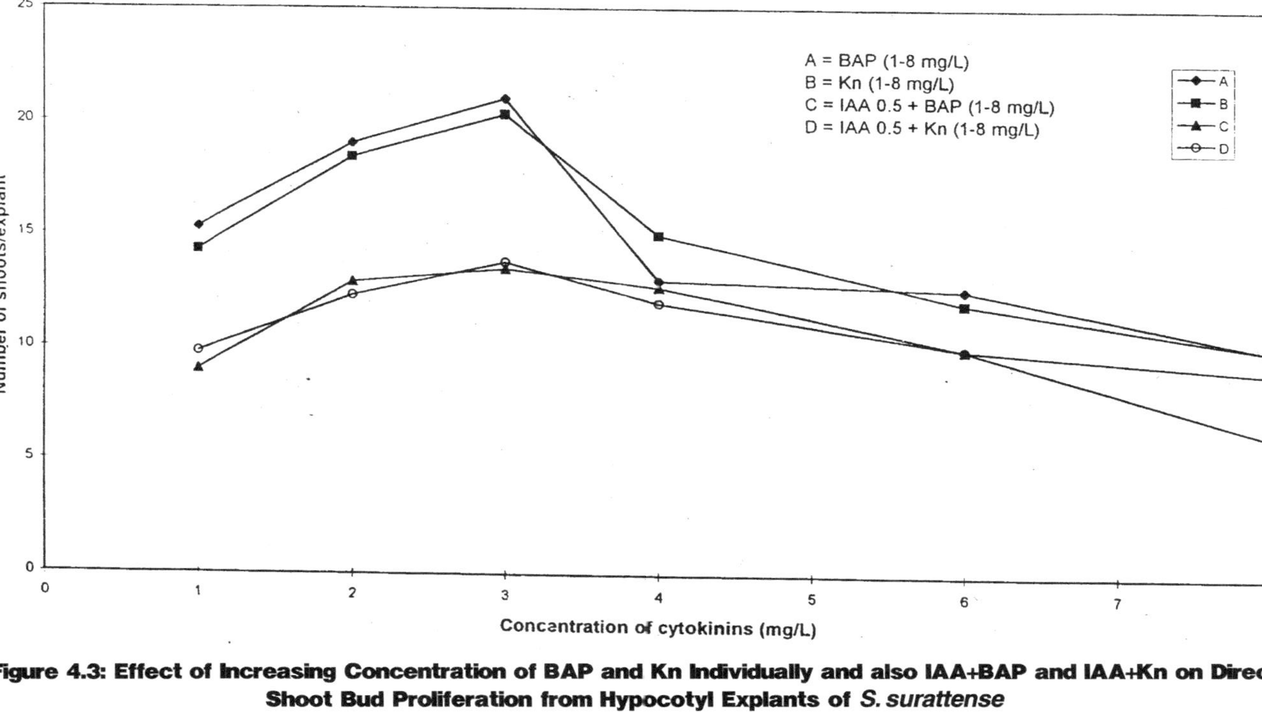

Figure 4.3: Effect of Increasing Concentration of BAP and Kn Individually and also IAA+BAP and IAA+Kn on Direct Shoot Bud Proliferation from Hypocotyl Explants of *S. surattense*

No difference was found in the induction ability of both the cytokinins used in hypocotyl explants. It was also found that even after addition of auxin to cytokinins the induction ability was not enhanced (Figure 4.3). This may be dependent on the endogenous hormonal level and exogenous supply to the explant.

Leaf explants

Leaf explants were cultured on MS medium containing different concentrations of cytokinins (BAP/Kn) alone and also in combination with (0.5/1.0 mg/L) IAA.

Leaf explants cultured on MS medium fortified with BAP/Kn as a sole growth regulator showed the direct organogenesis/shoot bud induction. All the cultures were responded (100 per cent) at 3.0 mg/L BAP but gradually the percentage of response was decreased upto 80.3 per cent. More number of shoots per explant (25.0 ± 0.35) was observed at 3.0 mg/L BAP and the number of shoot bud induction was found to be decreased as the concentration of BAP increased.

Leaf explants were cultured on MS medium amended with various concentrations of Kn to findout the difference between BAP and Kn in inducing the direct plant regeneration from leaf explants in *S. surattense*. Maximum number of shoot bud proliferation (11.3 shoots/explant) was found at 3.0 mg/L Kn compared to all other concentrations of Kn. Number of shoots/explant was reduced at high concentrations of Kn *i.e.*, at 4, 6 and 8.0 mg/L Kn as it was noted in BAP. Percentage of response was gradually increased upto 3.0 mg/L Kn and afterwards decreased the percentage of responding cultures. Likewise, the average number of shoots/explant development was increased gradually from 1.0 mg/L to 3.0 mg/L Kn but later it was decreased by inducing only 2.4 shoots/explant at high concentration (8.0 mg/L) of Kn. Similar results were also recorded on MS + BAP. However, high induction ability was found in all the concentrations of BAP compared to Kn (Figure 4.4).

To findout the influence of auxin-cytokinin combination on direct regeneration, the leaf explants were cultured on MS medium fortified with 0.5/1.0 mg/L IAA and different concentrations of cytokinins such as BAP/Kn in *S. surattense*. Leaf explants cultured on MS medium containing 0.5 mg/L IAA in combination with various concentrations of BAP, showed variable results (Figure 4.4).

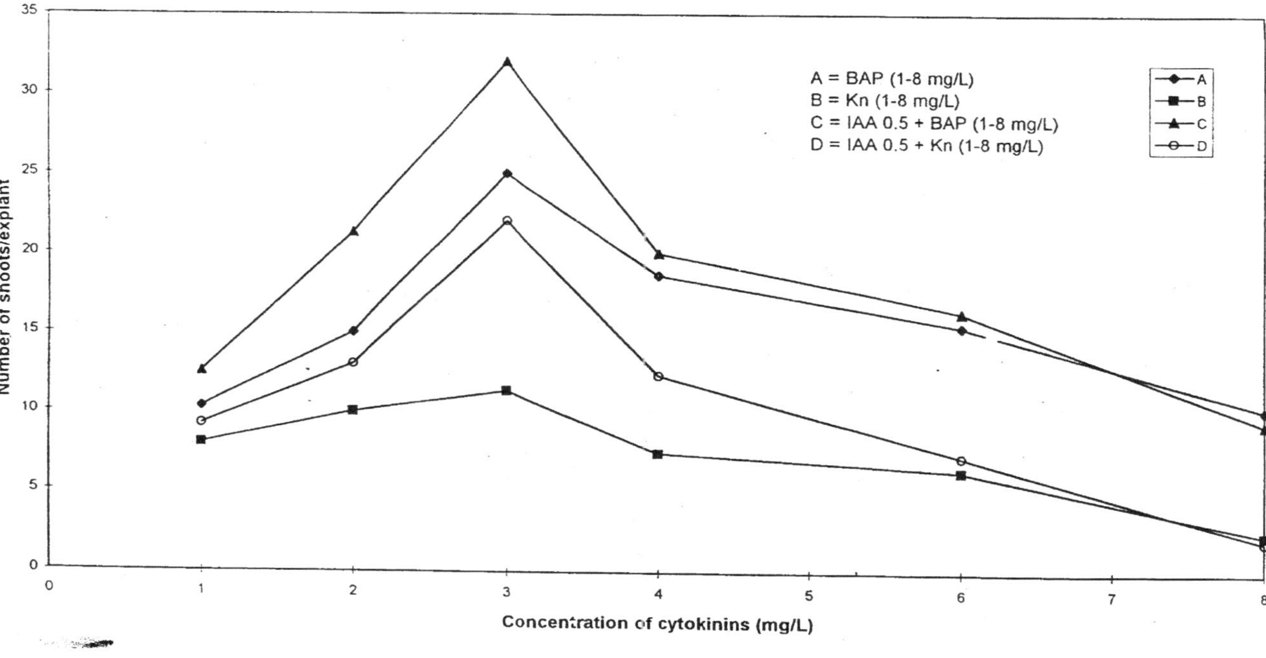

Figure 4.4: Effect of Increasing Concentration of BAP and Kn Individually and also IAA+BAP and IAA+Kn on Direct Shoot Bud Proliferation from Leaf Explants of *S. surattense*

Highest percentage of response was observed at 0.5 mg/L IAA + 3.0 mg/L BAP (Plate 3c). The percentage of response was increased upto 3.0 mg/L BAP and later gradually decreased at high concentrations. High frequency of shoots (32) was induced per explant at 3.0 mg/L BAP and the induction ability was decreased as the concentration of BAP incresed. When the concentration of auxin IAA increased to 1.0 mg/L along with BAP, the direct shoot bud regeneration efficiency was found to be decreased in all the concentrations of BAP tested. However, more number of shoots/ explant was induced at 3.0 mg/L BAP with 1.0 mg/L IAA (Plate 3d) but comparatively lesser than that of 0.5 mg/L IAA + 3.0 mg/L BAP combination.

To observe the difference in the direct organogenesis efficacy between BAP and Kn, the leaf explants were also cultured on 0.5/ 1.0 mg/L IAA in combination with various concentrations of Kn. Leaf explants cultured on MS medium augmented with 0.5 mg/L IAA and different concentrations of Kn showed maximum shoot buds induction (22.0 ± 0.5/explant) at 3.0 mg/L Kn with high percentage of responding cultures.

When the concentration of Kn has been increased beyond 3.0 mg/L, the average number of shoots/explant was gradually reduced (Figure 4.4). At high concentration of Kn (8.0 mg/L) only 2.0 ± 0.5 shoots/explant were induced. Likewise, the percentage of responding cultures was also found to be high (98.2) at 3.0 mg/L Kn and gradually it was decreased as the concentration of Kn increased. To findout the induction ability of direct shoot regeneration, the auxin concentration was increased to 1.0 mg/L IAA. Leaf explants cultured on MS medium fortified with 1.0 mg/L IAA and different concentrations of Kn (1-8 mg/L) showed the highest number of shoots induction at 3.0 mg/L Kn with 92.5 per cent of responding cultures Gradually the shoot regeneration was observed to be increased upto 3.0 mg/L Kn and after that decreased as the concentration was increased as observed in BAP. When the auxin concentration was increased to 1.0 mg/L IAA along with the cytokinin Kn it was found that the shoot regeneration was decreased in all the concentrations of Kn tested compared to 0.5 mg/L IAA.

On the whole, BAP showed superiority over Kn. It was also interesting to note that auxin-cytokinin combination produced more number of shoots compared to cytokinin as a sole growth regulator (Figure 4.4). The optimum level of IAA has been found to be 0.5 mg/L along with both the cytokinins used, because it induced the highest shoot bud regeneration capacity in comparison to 1.0 mg/L IAA.

Rooting of shoots

Microshoots (3-4 cm) developed from leaf regenerants were cultured on ½ strength MS medium supplemented with 0.5 mg/L and 1.0 mg/L IAA. Profuse rhizogenesis was observed on 0.5 mg/L (Plate 3e) compared to 1.0 mg/L IAA. 50 per cent of the regenerated shoots produced well developed roots (5-10) on MS medium containing 0.5 mg/L IAA whereas 30 per cent of shoots produced roots with 2-3 roots on 1.0 mg/L IAA.

Acclimatization

Rooted plantlets were transplanted to plastic pots containing vermiculite and perlite (3 : 1) and grown in culture room. The plantlets grew more and high percentage of plant survival (75 per cent) was achieved. After two weeks, the plants were transferred to soil successfully and grown in green house. The plants were found to be normal.

Plantlets were regenerated successfully from callus cultures on MS medium fortified with different concentrations of cytokinins *i.e.,* BAP/Kn individually and also in combination with 0.5 mg/L IAA/ 0.5 mg/L NAA. Maximum number of shoot buds were induced at 3.0 mg/L BAP in comparison to Kn as a sole growth regulator. When low level of auxins (0.5 mg/L) were added to the medium containing BAP/Kn, it was interesting to findout that the shoots induction was enhanced in all the concentrations of cytokinins tested. However, the shoot bud proliferation was found to be more on 0.5 mg/L IAA in combination with BAP/Kn compared to 0.5 mg/L NAA. Probably, IAA might have triggered the action of BAP/Kn in a proper way for inducing more number of shoots per explant. But the combination of IAA + BAP induced highest number of plantlet regeneration among all hormonal combinations and concentrations used.

Similarly, Hoque *et al.* (2000) have reported the high frequency of plant regeneration on MS medium containing 2.0 mg/L BA in combination with 0.5 mg/L NAA from cotyledon derived callus in *Momordica dioica*. They have also found the maximum number of shoots per explant on BA compared to Kn and induction was higher on NAA + BA than Kn as in *S. surattense* (Rama Swamy, *et al.*, 2006). Azad and Amin (1998) also found maximum number of shoot bud proliferation on the medium containing BA + NAA in internodal explants of *Adhatoda vasica* a medicinal plant. Similar observations were reported in callus cultures of *Theobroma cacao* (King and Rao, 1981) and *Piper longum* (Bhat *et al.*, 1992) the maximum proliferation of shoot buds on auxin and cytokinin combination as it was observed in *S. surattense.*

De Langhe and De Bruijne (1976) have observed the maximum shoot regeneration on BA + IAA in comparison to BA + NAA in leaf callus cultures of tomato. Similar differences in the organogenesis was also noted in *S. surattense.* Though, more number of shoots were formed per explant in auxin-cytokinin combination but the combination of IAA + BAP/Kn was more effective than NAA + BAP/Kn. Whereas Shahzad *et al.* (1999) have found the efficacy of auxin-cytokinin combination NAA + BAP in inducing shoot organogenesis from leaf callus cultures of *Solanum nigrum.*

In vitro micropropagation through callus cultures is an important technique to multiply and propagate a species in large numbers. But, the plantlets regenerated from callus cultures may show genetic variability and deviate from the normal diploids, which is reflected in various kinds of morphological abnormalities.

Therefore, direct shoot morphogenesis from the primary tissue (explant) is more desirable than via an intermediate callus phase (Larkin and Scrowcroft, 1981).

The explants such as cotyledon, hypocotyl and leaf cultured on MS medium fortified with different concentrations of cytokinins BAP/Kn alone and in combination with 0.5 mg/L and 1.0 mg/L IAA showed the direct shoot regeneration, whereas at 10 mg/L BAP/Kn alone and also in combination with 0.5 mg/L/1.0 mg/L IAA, the shoot bud induction was totally inhibited. Maximum number of shoots per explant was found at 3.0 mg/L BAP/Kn as a sole growth regulator in all the explants studied. When 0.5 mg/L/1.0 mg/L IAA added to the MS medium containing BAP/Kn showed the enhanced shoot bud proliferation in leaf and cotyledon explants with the exception of hypocotyl explants. But, the induction of shoot bud proliferation was decreased when the concentration of auxin IAA increased to 1.0 mg/L in all the explants and concentrations of BAP/

Kn tested. High frequency of shoots was induced on MS medium supplemented with 0.5 mg/L IAA + 3.0 mg/L BAP. Maximum shoots were developed per explant in cotyledon explants than leaf and hypocotyl explants. Less number of shoot bud proliferation was observed in hypocotyl explants. Thus, from the results it is evident that BAP showed superiority over Kn in inducing organogenesis directly from all the explants in *S. surattense* (see Figures 4.2-4.6). The same findings were recorded in *Capsicum* sp. (Philips and Hubstenberger, 1985).

The explants cotyledon, hypocotyl and leaf cultured on MS medium supplemented with IAA + BAP combination was found to be more effective in inducing maximum number of shoots in *Capsicum* spp. (Gunay and Rao,1978; Fari and Czako, 1981). Similarly, it was also reported in *Capsicum annuum* (Christopher and Rajam, 1996) and *Solanum melongena* (Sharma and Rajam, 1995). The combination of IAA + BAP showed superiority over IAA + Kn in all the explants in *S. surattense*. Similar findings were also made in *Petunia* (Rao and Harada, 1974), tomato (Kartha *et al.*, 1976) and *Physalis* (Bapat and Rao, 1977) and in *Solanum melongena* (Sadanandam and Farooqui, 1991).

Shahzad *et al.* (1999) have also observed the highest frequency of direct shoot regeneration on lower level of auxin and high level of cytokinin (4 mg/L BAP + 1 mg/L NAA) in leaf explants of *Solanum nigrum*. Highest number of shoots per explant were developed on MS medium containing 0.5 mg/L IAA + 2.5 mg/L BAP in leaf explants of *Solanum sisymbrifolium* (Rao *et al.*, 1997) compared to all other concentrations of BAP alone and also in combination with 1.0 mg/L IAA. When IAA concentration was increased to 1.0 mg/L, the shoot bud induction efficiency was reduced in *S. sisymbrifolium* similar to *S. surattense*.

Whereas Sudershan (1998) found the efficient adventitious shoot bud regeneration in leaf cultures cultured on MS medium supplemented with BA than NAA + BA in *Enicostemma axillare* a medicinal plant. These results are in contrast to the present observations in which lower level of auxin in combination with cytokinin enhanced the regeneration efficiency.

Direct shoot regeneration occurred on MS medium containing BAP/Kn and in combination with IAA in hypocotyl, stem and leaf cultures of *Solanum viarum* (Tejavathi and Bhuvana, 1998). They have also observed the maximum shoot regeneration on IAA + BAP, 2-ip and superiority of BAP over Kn. Jas Rai *et al.* (1999) reported the induction of large number of shoot buds from leaf disc explants of *Passiflora caerulea* when cultured on MS medium fortified with BA + IAA.

Seetharam *et al.,* (2003) have reported the essentiality of both cytokinin (BAP + Kn) combination for induction of multiple shoots from leaf cultures in *S. surattense.* In contrary to this, in our observations even sole cytokinin also induced the adventitious shoots from all the explants.

Neeta *et al.,* (2001) have reported the direct plantlet regeneration from different explants *i.e.,* hypocotyl, epicotyl, cotyledon and leaves cultured on IAA + BA combination, but they found the highest average number of shoot buds per leaf explant in neem. Similarly, it was also reported in *Capsicum* (Christopher and Rajam, 1996) and 7 genotypes of *Capsicum annuum* (Venkataiah and Subhash, 2001) that leaf explants produced more adventitious shoots followed by cotyledon and hypocotyl explants. Whereas the cotyledon cultures showed the efficiency in inducing high frequency of shoot buds proliferation in *S. surattense* (Figures 4.5 and 4.6).

Thus, the regeneration through callus mediated/direct organogenesis may vary from species to species. Even, the different explants cultured on the same level of cytokinins and auxin-cytokinin combinations also show varying results in the same species varieties. Morphogenic response of explants depend on many factors including balancing of exo–and endogenous auxin-cytokinin levels (Venkataiah and Subhash, 2001). Effect of these growth regulators on morphogenesis *in vitro* was well cocumented (Rao *et al.,* 1989). Endogenous hormones are triggered in explants by exogenous supply which interact with each other and leads to the establishment to some sort of balance to achieve regeneration (Vasil and Nitsch, 1975). Thus, the morphogenetic expression in a particular direction is manifested as a result of the cumulative effect of many factors *viz.,* growth regulators, nutrient medium, temperature, humidity, photoperiod etc. These are the important morphogenic tools which can control the internal milieu of the cell.

From the foregoing, it is evident that cotyledon explants were found to be more potential in producing high frequency number of shoots among all other explants tested in *S. surattense.* Cytokinins

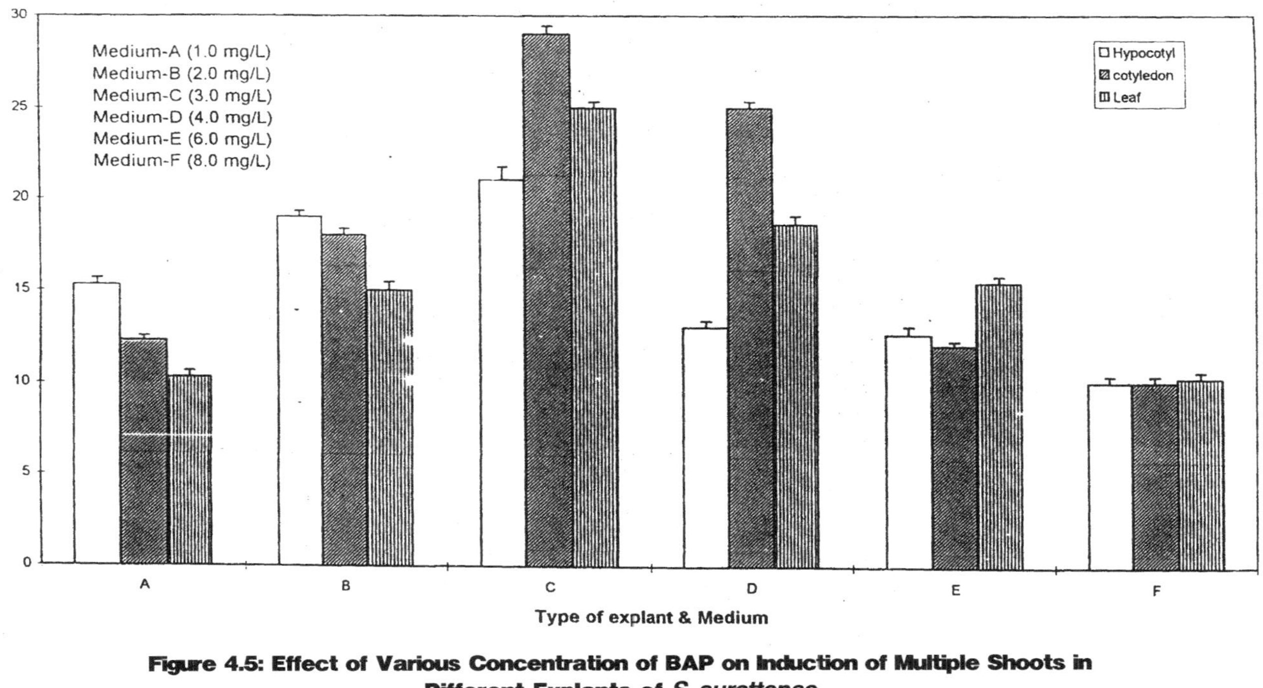

Figure 4.5: Effect of Various Concentration of BAP on Induction of Multiple Shoots in Different Explants of *S. surattense*

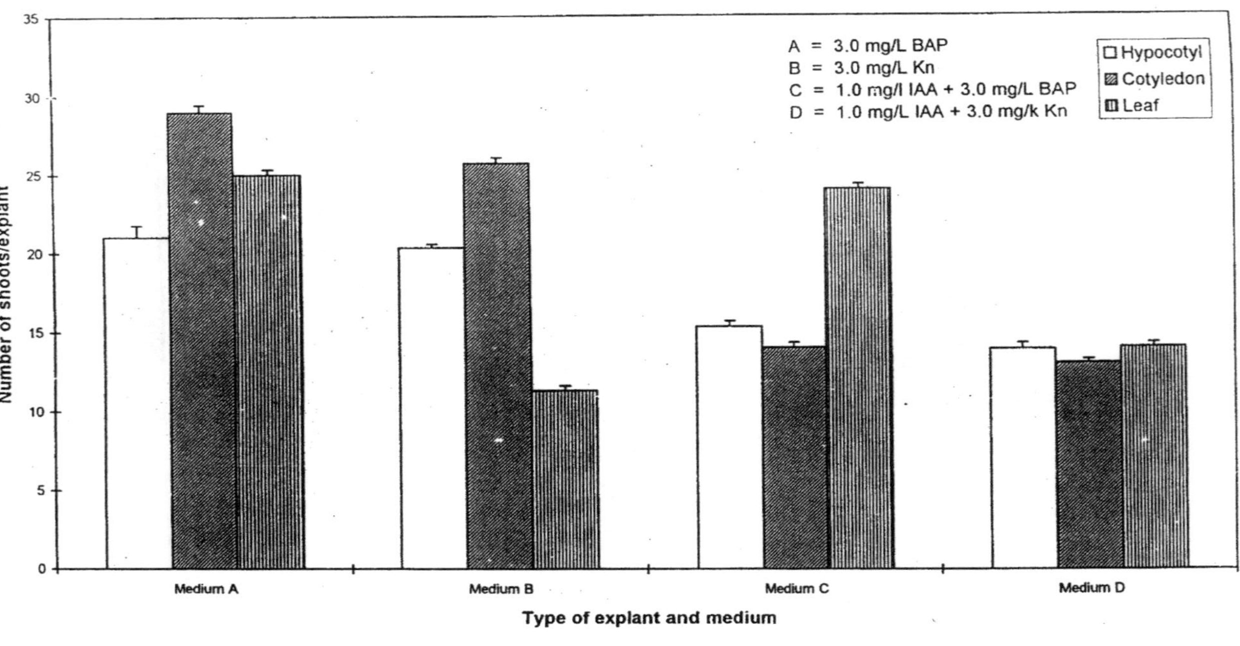

Figure 4.6: Effect of Auxins and Cytokinins on Maximum Number of Shoots Induction in Hypocotyl, Cotyledon and Leaf Explants of *S. surattense*

BAP/Kn alone or in combination with IAA was effective in inducing shoot regeneration in all the explants of *S. surattense*. However, 3.0 mg/L BAP/Kn with 0.5 mg/L IAA combination induced highest number of shoots except in hypocotyl explants. Thus, the plants regenerated *in vitro* by direct organogenesis may exhibit greater genetic stability than those produced from callus. The regeneration protocols developed from hypocotyl, cotyledon and leaf explants in the present investigation can be used for mass propagation of the species and also for genetic manipulation studies to introduce agronomically important traits.

Chapter 5

Somatic Embryogenesis

Somatic embryogenesis offers great potential in plant multiplication and crop improvement for efficient cloning and genetic transformation (Ammirato, 1987; Roberts *et al.*, 1995). It is an alternative and efficient method for plant propagation over regeneration via organogenesis.

Somatic embryogenesis is a process of formation of non-zygotic embryos from somatic cells.Somatic embryos are believed to originate from single cell while organogenesis is through collective organisation of cells. Therefore the plants derived from somatic embryos tend to be genetically alike; in addition somatic embryos are bipolar structures with root and shoot apices. Somatic embryos are developed in two ways (a) Direct embryogenesis *i.e.* without callus formation (b) Indirect embryogenesis (Reinert 1959) *i.e.* callus mediated. Somatic embryogenesis is initiated by (a) pre-embryogenic determined cells-PEDC (b) induced embryogenic determined cells-IEDC.

In recent years, somatic embryos are being used for developing synthetic seeds/artificial seeds for rapid and large-scale *in vitro* multiplication of elite and desirable plant species. These are also used for mutagenic studies and to raise the disease-free plants. The formation of plantlets from somatic embryogenesis is the ideal method of plant propagation for developing true-to-type of plants.

Following the initial reports of Reinert (1958) and Steward *et al.* (1958) the phenomenon of somatic embryogenesis has been achieved in a number of dicotyledonous plants (Banerjee and Gupta, 1975; Litz *et al.*, 1983; Price and Smith, 1984; Mujib *et al.*, 1990; Tewari *et al.*, 1999; Chand and Singh, 2001; Hussain *et al.*, 2000; Cardoza and D'Souza, 2000; Mohan *et al.*, 2000).

Somatic embryogenesis was reported from different explants in *Solanum melongena* (Gleddie *et al.*, 1983; Kalloo, 1993; Yadav and Rajam, 1997, 1998). Somatic embryogenesis has also been reported in medicinal plants: *Atropa belladona* (Thomas and Street, 1970), *Carum carvi* (Ammirato, 1977), *Papaver somniferum* (Nesslar, 1982), *Citrus limon* (Ben-Hayyim and Neumann, 1983), *Tylophora indica* (Mhatre *et al.*, 1984), *Chicorium intybus* (Heirwegh *et al.*, 1985), *Urginea indica* (Jha, 1986), *Cassia fistula* (Bajaj, 1988), *Tribulus terrestris* (Mohan *et al.*, 2000) and *Psoralea corylifolia* (Sahrawat and Chand, 2001).

Recently, Ramaswamy *et al.*, (2005b) have developed the protocol for plantlet regeneration through somatic embryogenesis in *S. surattense* a medicinally important plant.

Cotyledons and tender leaves cultured on various concentrations of NAA in combination with 0.5 mg/l BAP became swollen, and generally dedifferentiated and developed friable callus after 8-10 days of culture. Within 15-20 days of culture, globular embryos had formed directly on the surface of callus. When the explants of primary somatic embryos were cut into fragments and cultured on the same induction medium secondary somatic embryos were induced within two weeks.

Thus, proliferation of somatic embryos occurred in two ways : 1. Multiplication of somatic embryos from the explant through primary somatic embryogenesis; and 2. Proliferation of secondary somatic embryos from already formed somatic embryos through repetitive embryogenesis.

Among the various concentrations of NAA tested in combination with 0.5 mg/L BAP, the percentage of somatic embryo formation was found to be higher at 6.0 mg/L NAA + 0.5 mg/L BAP in cotyledon explants (Table 5.1; Figure 5.1). There was a generally increased tendency of somatic embryos formation with the increasing concentration of NAA in combination with BAP was observed. However, when the concentration of NAA increased to 10 mg/L + 0.5 mg/L BAP somatic embryogenesis was inhibited.

TABLE 5.1
Effect of Various Concentrations of NAA and 0.5 mg/L BAP on Somatic Embryogenesis in Cotyledon Explants of *S. surattense*

Growth Regulators (mg/L)			% of Cultures Responding	Percentage of Response for Somatic Embryo Formation	Average Number of Somatic Embryos/Explant (S.E.)*
NAA	+	BAP			
0.5	+	0.5	83.3	31.0	5.7 ± 0.32
1.0	+	0.5	86.0	50.7	6.9 ± 0.23
2.0	+	0.5	90.2	60.2	15.8 ± 0.27
3.0	+	0.5	91.2	76.2	20.0 ± 0.23
4.0	+	0.5	94.0	80.4	22.0 ± 0.35
6.0	+	0.5	90.7	85.0	24.0 ± 0.27
8.0	+	0.5	80.5	76.2	15.0 ± 0.37
10.0	+	0.5	70.5	—	—

*: Mean; ±: Standard Error.

Somatic embryogenesis was also induced from leaf explants on MS medium fortified with various concentrations of NAA (0.5–10 mg/L) in combination with 0.5 mg/L BAP (Table 5.2; Figure 5.1). Maximum number of somatic embryos/explant and higher percentage of response for somatic embryos formation have been found at 4.0 mg/L NAA + 0.5 mg/L BAP in leaf explants of *S. surattense* (Plate 4a-c). With the increase of NAA concentration upto 4.0 mg/L with 0.5mg/L BAP, there is a gradual enhanced somatic embryos induction was recorded. But, when the concentration of NAA was increased above 4.0 mg/L, percentage of response and somatic embryo induction were decreased. It was found that at higher concentration of NAA (10 mg/L) in combination with 0.5 mg/L BAP, the induction of somatic embryogenesis was inhibited as it was observed the same response in cotyledon explants. This might be due to altered hormonal levels in the medium which are critical for embryo formation. Leaf explants showed maximum percentage of somatic embryogenesis and high frequency of somatic embryo induction/explant (28.0 ± 0.27) compared to cotyledon explants (24.0 ± 0.25).

The calli developed from leaf and cotyledon explants containing globular embryos were transferred to maturation medium containing MS medium supplemented with 4.0 mg/L/6.0 mg/L NAA + 0.5 mg/L BAP respectively. Individual embryos enlarged into distinct bipolar structures and passed through each of the typical developmental stages (Globular, heart, torpedo and cotyledonary) (Plate 4b-d) after 4 weeks of culture. When these embryos with different developmental stages transferred to the same medium, further germination of embryos was not observed (Ramaswamy *et al.*, 2005b).

TABLE 5.2
Effect of Various Concentrations of NAA and 0.5 mg/L BAP on Somatic Embryogenesis in Leaf Explants of *S. surattense*

Growth Regulators (mg/L) NAA + BAP	% of Cultures Responding	Percentage of Response for Somatic Embryo Formation	Average Number of Somatic Embryos/Explant (S.E.)*
0.5 + 0.5	83	25.4	8.7 ± 0.23
1.0 + 0.5	86	56.0	9.9 ± 0.35
2.0 + 0.5	90	72.0	15.8 ± 0.73
3.0 + 0.5	92	80.0	25.3 ± 0.32
4.0 + 0.5	95	87.0	28.0 ± 0.27
6.0 + 0.5	98	75.0	25.0 ± 0.35
8.0 + 0.5	75	60.8	20.0 ± 0.32
10.0 + 0.5	70	—	—

*: Mean; ±: Standard Error.

Hence, the somatic embryos with various developmental stages (heart, torpedo and cotyledonary) were further subcultured on fresh MS medium containing various concentrations of BAP (1.0–8.0 mg/L) in combination with 0.5 mg/L IAA for germination of somatic embryos induced from cotyledon and leaf explants. Of these media tested, MS + 0.5 mg/L IAA + 2.0 mg/L BAP proved to be the best for somatic embryo germination and plantlet formation after 3 weeks of culture.

A major factor for somatic embryogenesis is the nature of growth regulators used in the induction medium. The type of auxin or auxin in combination with cytokinin used in the medium can greatly influence somatic embryo frequency. The requirement of cytokinin

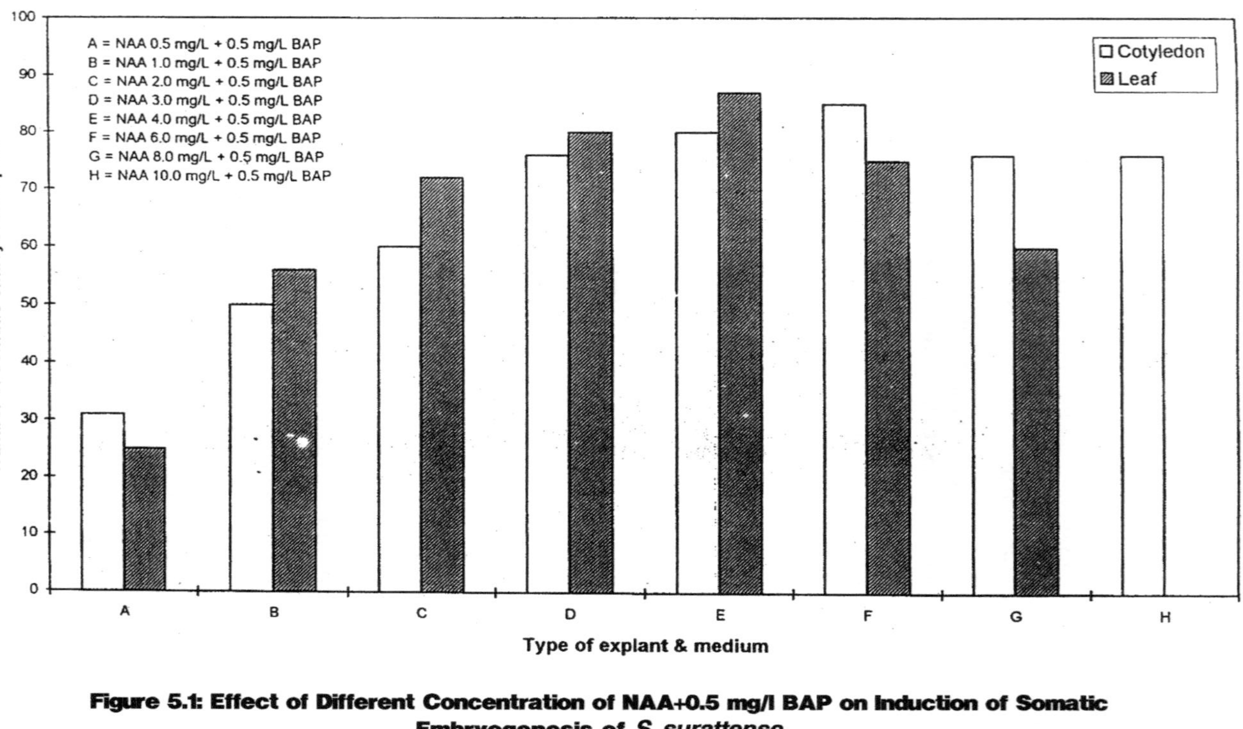

Figure 5.1: Effect of Different Concentration of NAA+0.5 mg/l BAP on Induction of Somatic Embryogenesis of *S. surattense*

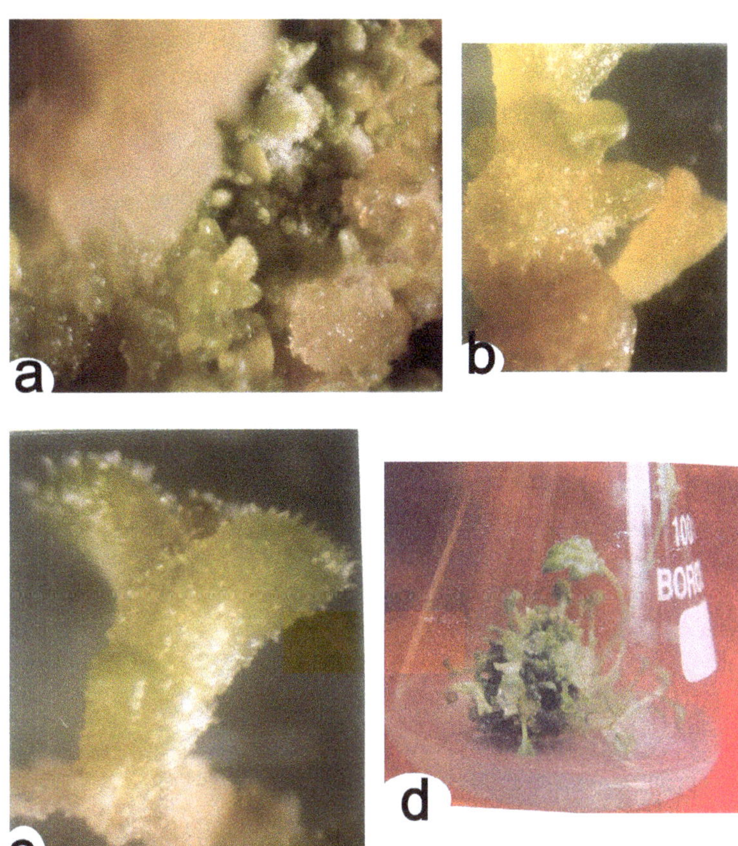

Plate 4a: Embryogenic Callus Showing Various Stages of Somatic Embryoids on MS + 4.0 mg/L NAA + 0.5 mg/L BAP after 3 Weeks of Culture in *S. surattense*

Plate 4b: Heart-shaped Somatic Embryo on MS + 4.0 mg/L NAA + 0.5 mg/L BAP

Plate 4c: Cotyledonary Stage Somatic Embryo on MS + 6.0 mg/L NAA + 0.5 mg/L BAP

Plate 4d: Conversion of Cotyledonary Stage Embryos into Plantlets on MS + 0.5 mg/L IAA + 2.0 mg/L BAP

in addition to auxin was observed in *Sapindus trifoliatus* (Desai *et al.*, 1986), *Terminalia arjuna* (Kumari *et al.*, 1998), and *Psoralea corylifolia* (Sahrawat *et al.*, 2001), as it was observed in *S. surattense*. While Reddy and Reddy (1993) have reported the improved response of auxin 2,4-D alone for induction of somatic embryogenesis compared to 2,4-D and cytokinin combinations in *Arachis hypogaea*. Somatic embryogenesis was induced on MS medium containing NAA alone in *Solanum melongena*. However its optimal concentration varies with the initial explant. Leaf explants required 2-6 mg/L NAA for somatic embryo differentiation while hypocotyls required 6-10 mg/L (Matsuoka and Hinata, 1979; Gleddie *et al.*, 1983; Sharma and Rajam, 1995).

TABLE 5.3
Effect of IAA + BAP on Germination of Somatic Embryo in *S. surattense*

Medium + Growth Regulators mg/L	Germination Frequency (± S.E.) *
½ MSO	--
MSO	--
MS + IAA (0.5) + BAP (1.0)	31.6 ± 0.56
MS + IAA (0.5) + BAP (2.0)	71.2 ± 1.21
MS + IAA (0.5) + BAP (3.0)	63.0 ± 0.72
MS + IAA (0.5) + BAP (4.0)	35.5 ± 0.31
MS + IAA (0.5) + BAP (6.0)	23.0 ± 0.25
MS + IAA (0.5) + BAP (8.0)	11.3 ± 1.30

* Data represents mean ± standard error of three experiments each consisting of 20 somatic embryos.

Direct somatic embryogenesis was reported by adding BAP to the medium and also the number of embryos further increased by enriching the medium with NAA in *Hippeastrum hybridum* (Mujib *et al.*, 1998) Similar findings were also made by Cavallini and Natali (1989) in *Brimeura amethystina*. Sahrawat *et al.* (2001) have also observed the high-frequency somatic embryogenesis in hypocotyl explants on MS medium supplemented with NAA (1.4 mM) + BAP (2.2 mM). Setterfield (1963) opined that auxins and cytokinins may act synergistically to promote both cell division and cell expansions depending upon other factors within the cell which react with these hormones.

Somatic embryogenesis is generally believed to be triggered by an auxin and for many plants, 2,4-D has been widely regarded to be effective for somatic embryogenesis (Litz *et al.*, 1982; Ammirato, 1983 a,b; Finner, 1988; Baker and Wetzstein, 1994). Chand and Singh (2001) have recorded the induction of somatic embryogenesis using 2,4-D in *Hardwickia binata*. Similarly, 2,4-D induced the somatic embryogenesis in *Mangifera indica* (Hussain *et al.*, 2000). Despite the crucial role of 2,4-D in inducing somatic embryogenesis its continuous presence of auxin inhibited the embryo development into cotyledonary stage in both Amarapali and Chausa cvs. of mango (Hussain *et al.*, 2000). According to Zimmerman (1993) new gene products are needed for the progression from the globular to the heart-stage and these new products are synthesized only when an exogenous auxin is removed. But, for morphogenesis of somatic embryos in *S. surattense*, auxins and cytokinin combination is required. At higher concentration of auxin probably the population of embryogenic cells drops due to their disruption and elongation and the embryogenic potential of the culture is lost (Bhojwani and Razdan, 1996). Similarly, in *S. surattense*, embryogenesis was inhibited at 10 mg/L concentration of NAA. Sinha *et al.* (2000) have found the efficacy of Kn in inducing somatic embryogenesis in leaf explants of *Sapindus mukorossi*.

Maturation process is a critical step in somatic embryogenesis which leads to the complete plantlet formation. In the present investigation, both auxin and cytokinin combination favoured the maturation and germination of somatic embryos. Rao and Lakshmi Sita (1996) have observed the maturation and conversion to whole plants from somatic embryos cultured on MS medium supplemented with cytokinin BAP (0.5–2.0 mg/l). The same results were made in *Pharbitis nil* (Jia and Chna, 1992) and water melon (Compton and Gray, 1993). In other species, the importance of ABA and mannitol for differentiation and further germination of somatic embryos was observed.

Whereas Binzel *et al.* (1996) reported that the entire process of induction and maturation of the embryos was completed on the same MS medium containing auxins and cytokinins (2,4-D + TDZ) in *Capsicum annuum* as it was observed the requirement of both the hormones in the present investigations. Similarly, somatic embryos maturation on MS medium containing the combination of auxins (NAA) and cytokinins (BAP) was observed in *Prunus avium* (Garin *et al.*, 1997) and *Hardwickia binata* (Chand and Singh, 2001),.

Thus, somatic embryogenesis always appeared to be dependent on the type of auxin/cytokinin/auxin + cytokinin and their concentration in the medium. The type of phytohormone and its concentration also varies from genotype to genotype. High concentration of auxin in combination with less concentration of cytokinin induced the somatic embryogenesis and maturation of somatic embryos in *S. surattense*. However, for germination of somatic embryos, low level of auxins and high concentration of cytokinin combination is required.

Regeneration via embryogenesis is better for obtaining genetically uniform plants than through organogenesis. It is evident that the somatic embryogenesis in this species will be useful in the improvement of this medicinally important species (Rama Swamy *et al.*, 2005b). Somatic embryogenesis is also preferred because it allows production of plants without somaclonal variation and in efficient cloning and genetic transformation. Synthetic seeds can also be developed by encapsulating somatic embryos in sodium alginate complexed with calcium chloride as it was developed in *S.melongena* (Lakshmana Rao and Singh, 1991; Mariani, 1992).

Chapter 6

Micro-propagation/Clonal Propagation

Tissue culture techniques are now being used for effective plant propagation of several horticultural and agricultural crops. Micropropagation is a relatively new technology and application of this method would serve to overcome barriers to progress in the multiplication of elite genotypes. Micropropagation/clonal propagation aims at production of plants, breeding true-to-type, in larger numbers of shorter time and disease-free. It is an important *in vitro* technique for rapid and mass multiplication of identical and isogenic lines by involving shoot tip culture, adventive bud induction or by stimulation of axillary bud or lateral meristem.

Micropropagation generally involves three steps: (*i*) initiation of aseptic cultures (shoot tip/meristem tip/nodal or axillary bud); (*ii*) rooting of *in vitro* formed shoots; (*iii*) transplantation of regenerated plants into the field. This technique has been used for *in vitro* mass multiplicationof many a number of medicinally, commercially and economically important species and also conservation of endangered and threatened species : *Digitalis* (Maria *et al.*, 1990), *Eucalyptus* (Lakshmisita and Vaidyanathan, 1979; Grewal *et al.*, 1980), *Leucaena* (Dhawan and Bhojwani, 1985), *Prosopis*

(Goyal and Arya, 1984; Jordan *et al.*, 1985; Tabone *et al.*, 1986), *Tectona* (Lakshmisita and Chatopadhyaya, 1986), among the genus *Acacia: A. albida* (Duhoux and Davis, 1985), *A. mangium* (Bhaskar and Subhash, 1996), *Decalepis hamiltonii* (Bais *et al.*, 2000), *Canavalia virosa* (Kathiravan and Ignacimuthu, 1999), *Ocimum sanctum* (Anwar and Siddiqui, 2000), *Pisonia alba* (Jagadishchandra *et al.*, 1999), *Bacopa monniera* (Tiwari *et al.*, 2000) *Bixa orellana* (Sharon and D'Souza, 2000), *Zizyphus mauritiana* (Sudershan *et al.*, 2000).

SHOOT TIP CULTURE

Shoot regeneration from shoot apex/shoot tip is direct, relatively simple and is not prone to somaclonal variation and chromosomal abnormalities. This is an elegant methodology of multiplying plants *in vitro* starting from a single shoot with obvious potential when applied to crop plants (Quak, 1977).

Culture of shoot meristem, especially through enhanced branching, permits rapid clonal propagation and a high degree of genetic uniformity of the progeny (Sunitha and Handique, 2000). This mericlone technology has been wide spread practical application in producing virus-free plants *in vitro* in recent years (Bhaskaran *et al.*, 1992; Nasir *et al.*, 1997; Geetha and Shetty, 2000; Sharon and D'Souza, 2000; Sudershan *et al.*, 2000, Kaur *et al.*, 2000; Emmanuel *et al.*, 2000). Although mainly used for virus elimination, meristem-tip culture has also enabled plants to be freed from other pathogens, including viroids, mycoplasms, bacteria and fungi. Also, meristem freeze preservation as a method of conservation of germplasm has made possible to utilize when needed.

The latest technology for delivery of genes into plant tissues is the biolistic gun which requires regenerable tissues. So, target tissues may be callus, suspension cells, cotyledons, leaves, meristem tips or any other regenerable explant. Recently, shoot meristem tips have also been used for direct delivery of desired genes in soyabean, cotton and sorghum (Gould *et al.*, 1991; Bhaskaran *et al.*, 1992; Mc Cabe and Martinel, 1993). In view of the importance of mericlone technology in phytopathology and genetic engineering, experiments were conducted to achieve direct multiple shoot regeneration from shoot apex in *S. surattense* (Rama Swamy *et al.*, 2006b).

Direct multiple shoot proliferation was observed in shoot tip cultures on MS medium fortified with different concentrations of

BAP/Kn alone and also in combination with NAA in *S. surattense* (Tables 6.1 and 6.2). Callus formation was not observed in the above treatments. Number of shoots increased with increased levels of BAP. Of the various treatments tested on MS medium, BAP at 3.0 mg/L resulted in maximum number of shoots (8.5 ± 0.3) per explant that were supported by maximum length of shoots (16.3cms ± 1.6) and number of leaves compared to all other concentrations of BAP and Kn used (Plate 5C). Whereas maximum number of shoots (10.3 ± 1.2) were developed from the shoot tips, cultured on MS medium fortified with 0.1 mg/L NAA and 3.0 mg/L Kn (Plate 5a) followed by 0.1 mg/L NAA and 3.0 mg/L BAP and 0.5 mg/L NAA + 3.0 mg/L Kn.

TABLE 6.1
Effect of Different Concentrations of BAP and NAA + BAP on *In vitro* Multiple Shoot Formation from Shoot Tip Cultures of *S. surattense*

Growth Hormone (mg/L)		Average Number of Shoots/Explant (S.E.)*	Length of Shoots (cm) (S.E)*
NAA	BAP		
—	1.0	2.5 ± 0.3	10.3 ± 1.3
—	2.0	3.8 ± 0.2	10.2 ± 0.4
—	3.0	8.5 ± 0.3	16.3 ± 1.6
—	4.0	3.2 ± 0.2	9.4 ± 0.3
0.1	3.0	9.5 ± 1.2	8.5 ± 0.9
0.5	3.0	5.6 ± 0.6	12.3 ± 0.6
0.1	4.0	4.2 ± 0.5	10.6 ± 0.5
0.5	4.0	2.3 ± 0.4	13.0 ± 1.7

*: Mean; ±: Standard Error.

However, higher concentration of BAP (4.0 mg/L) either alone or with NAA (0.5 mg/L) on MS medium resulted in decrease in number of shoots irrespective of concentration of BAP in the medium. Considerable decrease in shoot length was observed but with the addition of NAA (0.1 mg/L) to BAP shoot length was increased with the number of shoots.

Morphogenetic response of shoot tip cultures on various concentrations of cytokinin such as Kn alone and in combination with NAA is presented in Table 6.2. As the concentration of Kn was increased, the number of shoots per explant was also produced more. But at high concentration of Kn (4.0 mg/L) considerably the number of shoot induction was found to be reduced. Whereas with the addition of NAA to Kn, maximum number of shoots/explant was developed (Plate 5a) with lengthy shoots in comparison to the shoot tips cultured on MS medium supplemented with Kn alone. However, maximum length of shoots were found at 0.1 mg/L NAA + 4.0 mg/L Kn among all other combinations and concentrations of NAA + BAP and NAA + Kn tested. Shoot tips cultured on MS basal medium developed an elongated single shoot. Phytohormones like IAA and IBA alone or in different combinations did not support shoot proliferation but caused callus or root formation.

TABLE 6.2

Effect of Different Concentrations of Kn and NAA + Kn on *In vitro* Multiple Shoot Induction from Shoot Tip Cultures of *S. surattense*

Growth Hormone (mg/L)		Average Number of Shoots/Explant (S.E.)*	Length of Shoots (cm) (S.E)*
NAA	Kn		
—	1.0	1.2 + 0.5	11.5 ± 1.3
—	2.0	3.5 ± 0.5	10.2 ± 0.3
—	3.0	6.5 ± 1.5	12.5 ± 1.3
—	4.0	2.2 ± 1.5	10.5 ± 0.3
0.1	3.0	10.3 ± 1.2	13.8 ± 1.0
0.5	3.0	8.5 ± 0.5	10.3 ± 1.3
0.1	4.0	7.3 ± 0.9	15.3 ± 0.5
0.5	4.0	3.3 ± 0.5	10.5 ± 0.5

*: Mean; ±: Standard Error.

For subculture, individual shoots grown on MS medium containing 0.1 mg/L NAA + 3.0 mg/L BAP/Kn were excised and subcultured on the fresh medium with the same combination of hormones. In each subculture, axillary and adventitious shoots grew from each shoot with lateral branches too.

Root induction ability of *in vitro* regenerated shoots from shoot tip cultures, cultured on rooting medium (RM) *i.e.*, ½ Strength MS medium supplemented with IAA and IBA is presented in Table 6.3.

TABLE 6.3
Rooting Ability of Regenerated Shoots from Shoot Tip of *S. surattense* Cultured on □ Strength MS Medium Supplemented with IAA and IBA

Growth Hormone (mg/L)		Percentage of Response	Average Number of Roots (S.E)*	Length of Roots (cm) (S.E)*
IAA	IBA			
0.0	0	20	1.0 ± 0.1	5.6 ± 0.2
0.5	—	60	2.2 ± 0.1	10.3 ± 0.2
1.0	—	80	8.8 ± 0.4	12.3 ± 0.5
—	0.1	40	1.5 ± 0.3	10.2 ± 0.5
—	0.5	80	8.6 ± 0.6	12.5 ± 1.3

*: Mean; ±: Standard Error.

RM stimulated root emergency from the cut ends of the shoots after 10–12 days. Single root was induced with short root on MS basal medium. Whereas RM containing 1.0 mg/L IAA and 0.5 mg/L IBA induced more number with lengthy roots (Plate 5b).

After four weeks of root induction, the plantlets were thoroughly washed with running tap water, planted in sterile soil mix containing, sand : peat moss and humus (1:1:1) and gradually acclimatized at normal culture room temperature for 25 days. Then these plantlets were transferred to green house. Thus hardened plantlets were maintained in the green house for the field transplantation (Figure 6.1).

Maximum number of shoots were induced on MS medium amended with the combination of both auxins and cytokinins in *S. surattense*. These results are also in agreement with those on *Tectona grandis* (Gupta *et al.*, 1980), *Albizzia lebbeck* (Gharyal and Maheshwari, 1982), *Mitragyna parviflora* (Roy *et al.*, 1988, 1996). Multiple shoot induction was also observed in *Zizyphus mauritiana* (Sudershan *et al.*, 2000) and *Vanilla planifolia* (Geetha and Shetty, 2000) shoot tips cultured on MS + cytokinin alone as it was observed in *S. surattense.*

Gupta *et al.*, (1997) reported the multiple shoot bud induction from shoot apex cultured on MS medium containing BA in cotton. Nasir et al, (1997) have studied the shoot meristem culture in 16 cultures of cotton using several media formulations. They observed the best shoot development on MS media containing Kn alone

Explant: shoot tips

Surface sterilization (0.1 per cent $HgCl_2$, 4-5 min)

Excised meristem tips from healthy shoots (Apical dome + one or more leaf primordia, 0.5–0.8 cm)

Culture on suitable medium with hormonal additives (BAP/Kn, NAA + BAP/Kn)

Plantlet Regeneration

Plantlets transferred to sterilized soil mix in plastic pots and kept in culture room

Transfer of plantlets to potted soil in green house for maintenance

Growth of virus-free plants

Figure 6.1: Flow Chart for Micropropagation through Shoot Meristem Culture of *S. surattense*

compared to other media with NAA/IAA in combination with Kn. These results are contrary to the observations made in *S. surattense* in which auxins in combination with cytokinins showed the increased number of shoots/explant. Sharma and Dhiman (1998) have also observed the similar results when they have cultured the shoot tips

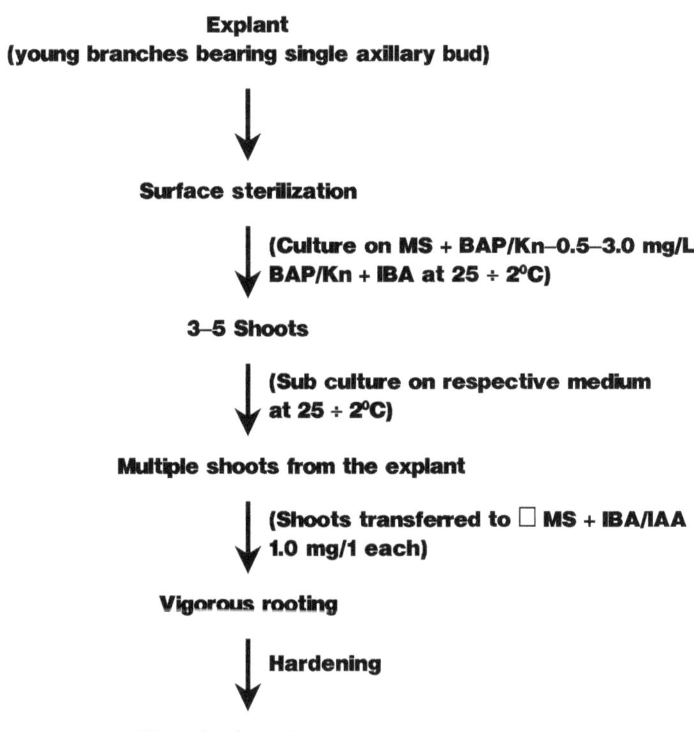

Figure 6.2: Flow Chart for Micropropagation through Axillary Bud Culture of *S. surattense*

of F1 hybrids of *Paulownia.* They found the highest multiple shoot induction on MS + NAA + BAP combination and less number of shoots on MS + BAP/Kn alone.

Zaman et al, (1996) have studied the effect of different cytokinins *viz.*, BA, Kn, 2-ip and zeatin on multiple shoot induction from shoot tip cultures in mulberry. According to their observations, BA and Kn were superior to 2-ip and zeatin. The superiority of BA over other cytokinins for multiple shoot formation has been reported in mulberry by Hossain et al, (1991) as it was observed in *S. surattense.*

Das and Mitra (1990) and Roy et al, (1993) have also reported the requirement of both auxin and cytokinin for induction of multiple

Plate 5a: Multiple Shoots Formation from Shoot Tip Culture on MS + 0.1 mg/L NAA + 3.0 mg/L Kn After 6 Weeks of Culture in *S. surattense*

Plate 5b: Proliferation of Shoots with Rooting from Shoot Tip Regenerated Plantlet on MS + 1.0 mg/L IAA

Plate 5c: Shoot Elongation from an Axillary Bud and Also Multiple Shoot Buds Proliferation on MS + 2.0 mg/L BAP

Plate 5d: Hardened Micropropagated Plant in a Plastic Pot

shoots in *Eucalyptus tereticornis* and Jack fruit respectively. Thus a combination of both auxin and cytokinin improved the efficiency of multiple shoots development, although it depended on the combination and concentrations employed. Whereas Pawar *et al.* (2002) have reported the induction of multiple shoots on MS medium fortified with BAP/Kn alone in *S. surattense* as observed in the present investigations.

Direct shoot bud proliferation/regeneration was found from the shoot apices without an intervening callus stage in *S. surattense* which is advantageous because explants could be taken from selected elite plants. The plants which are produced by direct organogenesis exhibit greater genetic stability than those produced from callus. Thus, the protocol developed for shoot and root development from meristematic shoot tips is quite simple and needs less time and labour to regenerate large number of plantlets. Hence, this regeneration system can be exploited for screening of large numbers of biolistic gun-transformed meristematic shoot tips for desirable genes as a model system.

NODAL CULTURE

The axillary/nodal bud induction is one of the most efficient method of micropropagation technique since the emerging buds, especially from meristematic organs and tissues possess a great potential for vigorous development (Yadav *et al.*, 1990; Pattanaik *et al.*, 1995). Axillary buds have been found to be most suitable for clonal propagation of medicinally important species: *Morus niger* (Yadav *et al.*, 1990), *Commiphora wightii* (Durga and Mehta, 1993), Mulberry (Pattanaik *et al.*, 1995), *Jasminum officinale* (Bhattacharya and Bhattacharya, 1997), *Ixora singaporensis* (Malathy and Pai, 1998), *Melia azedarach* (Raghuraman and Ramanujam, 1998), *Zizyphus mauritiana* (Sudershan *et al.*, 2000), *Decalepis hamiltonii* (Bais *et al.*, 2000), *Vanilla planifolia* (Geetha and Shetty, 2000), *Plumbago indica* (Sunitha and Handique, 2000), *Canavalia virosa* (Kathiravan and Ignacimuthu, 1999), *Ocimum sanctum* (Shahzad and Siddique, 2000), *Pisonia alba* (Jagdishchandra *et al.*, 1999), *Bacopa monniera* (Tiwari *et al.*, 2000) and *Bixa orellana* (Sharon and D'Souza, 2000).

The results on nodal culture of *S. surattense* cultured on different hormonal combinations and concentrations were presented in (Tables 6.4–6.6) and showed in Figures 6.3–6.6. The axillary buds became active within week after inoculation and new shoots became disctinct by the second and third week with leaves and internodes.

TABLE 6.4
Effect of Different Concentrations of BAP and IBA + BAP on Multiple Shoot Induction from Nodal Bud Cultures of *S. surattense*

Growth Hormone (mg/L)		Percentage of Response	Average Number of Shoots/Explant (S.E)*	Length of Roots (cm) (S.E)*
IBA	BAP			
—	0.5	+	1.0 ± 0.5	5.0 ± 0.12
—	1.0	++	2.5 ± 0.8	8.5 ± 0.15
—	2.0	+++	3.8 ± 0.7	10.5 ± 0.5
—	3.0	++	2.4 ± 0.5	7.5 ± 0.2
0.1	2.0	+++	4.5 ± 0.7	8.5 ± 0.5
0.5	2.0	+++	4.0 ± 0.5	9.5 ± 0.2
0.1	3.0	++	3.5 ± 0.6	8.6 ± 0.5
0.5	3.0	++	2.8 ± 0.5	8.7 ± 0.7

+: below 50%; ++: Above 50%; +++: Above 75; * Mean; ±: Standard Error.

TABLE 6.5
Effect of Different Concentrations of Kn and IBA + Kn on Multiple Shoot Induction from Nodal Bud Cultures of *S. surattense*

Growth Hormone (mg/L)		Percentage of Response	Average Number of Shoots/Explant (S.E)*	Length of Roots (cm) (S.E)*
IBA	Kn			
—	0.5	+	2.5 ± 0.5	6.0 ± 0.5
—	1.0	++	3.6 ± 0.3	7.5 ± 0.3
—	2.0	+++	4.5 ± 0.6	8.4 ± 0.4
—	3.0	++	3.4 ± 0.5	6.5 ± 0.3
0.1	2.0	+++	6.5 ± 0.3	7.5 ± 0.5
0.5	2.0	+++	7.0 ± 0.5	8.6 ± 0.3
0.1	3.0	+++	4.5 ± 0.3	6.7 ± 0.5
0.5	3.0	++	4.6 ± 0.5	6.8 ± 0.7

+: below 50%; ++: Above 50%; +++: Above 75; * Mean; ±: Standard Error.

The explants survival from nodal segments of mature plant of *S. surattense* varied with season. Explants collected in August to

November period showed less time for sprouting and quick shoot bud proliferation and also recorded high percentage (60–70 per cent) of explants establishment. On the other hand very low percentage (30–50 per cent) of explants establishment was found during December to July period (Ramaswamy *et al.*, 2004).

TABLE 6.6
Rooting Ability of Regenerated Shoots from Nodal Cultures of *S. surattense* on □ Strength MS Medium with Various Concentrations of IAA and IBA

Growth Hormone (mg/L)		Percentage of Response	Average Number of Shoots/Explant (S.E)*	Length of Roots (cm) (S.E)*
IAA	IBA			
0.0	–	+	Single	5.3 ± 0.5
0.1	–	+++	2.0 ± 0.2	15.3 ± 0.4
0.5	–	+++	4.0 ± 0.1	20.4 ± 0.5
1.0	–	++	9.0 ± 0.5	25.4 ± 0.4
–	0.1	+++	2.1 ± 0.1	18.5 ± 0.5
–	0.5	+++	3.5 ± 0.2	21.5 ± 0.3
–	1.0	+++	7.8 ± 0.1	22.5 ± 0.3

+: below 50%; ++: Above 50%; +++: Above 75; * Mean; ±: Standard Error.

The size of the nodal explants was found to play an important role in initiation and elongation of shoots. The smaller (1cm) explants could initiate more multiples than the longer (2.0 cm) nodal explants. Although 2.5 cm long segments produced less number of shoot buds, they showed better elongation. Whereas the biggest nodal segment (3.0 cm) tried failed to give multiples, only a single shoot developed from each node.

The medium containing 2.0 mg/L BAP induced maximum number of shoots 3.8 ± 0.7 with longer shoots (10.5 ± 0.5 cm) and also showed high percentage (75 per cent) of responding cultures (Plate 5c). As the concentration of BAP was increased upto 2.0 mg/L gradually, the shoot bud proliferation was also found to be increased and when BAP concentration was increased above 2.0 mg/L, the rate of shoot multiplication and elongation was reduced (Table 6.4).

Addition of auxin such as IBA (0.1 mg/L) to cytokinin, BAP (2.0 mg/L) together in the medium produced maximum number (4.5 ± 0.7) of shoots with high percentage of responding cultures compared to all other combinations and concentrations tested. Whereas the medium containing 0.5 mg/L IBA with 2.0 and 3.0 mg/L BAP reduced the number of shoot bud induction and also showed the less percentage of explant establishment (Figures 6.4 and 6.6).

Maximum number of shoot bud proliferation (4.5 ± 0.6) with longer shoots was found on MS medium fortified with 2.0 mg/L Kn alone. High percentage of cultures were also responded on the same medium (Figures 6.3 and 6.5). As the concentration of Kn was increased in the medium, the percentage of responding cultures and the shoots formation/explant was also found to be increased gradually (Table 6.5). Whereas at high concentration of Kn (3.0 mg/L) less number of shoots was developed with decreased in length of shoots as it was observed in the case of BAP in the present investigations.

To find out the synergistic effect of both auxins and cytokinins together such as IBA (0.1 and 0.5 mg/L) was added to the medium containing Kn (2.0 and 3.0 mg/L). More number of shoots was

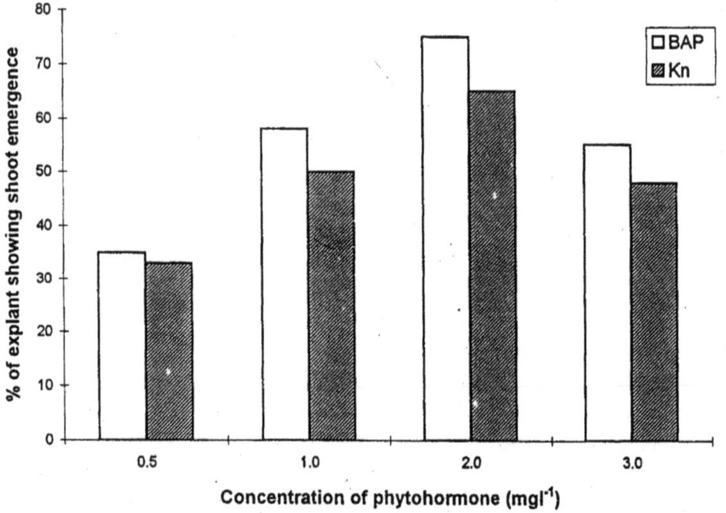

Figure 6.3: Effect of Different Concentrations of BAP/Kn on Nodal Culture in *S. surattense*

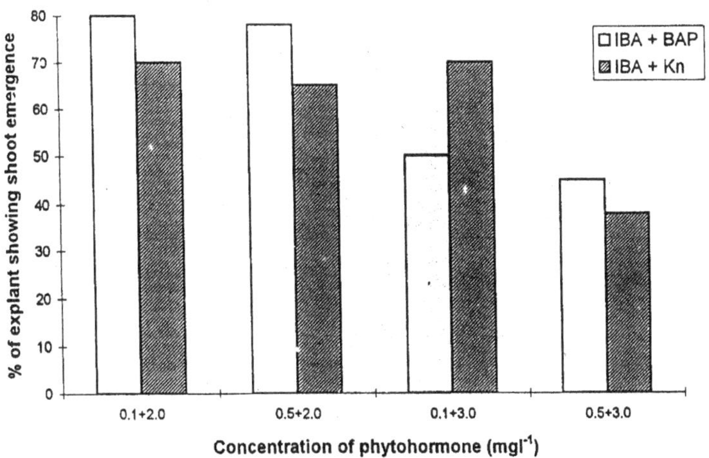

Figure 6.4: Effect of Different Concentrations of IBA+BAP/Kn on Nodal Culture in *S. surattense*

formed/explant and also with longer shoots in the medium fortified with 0.5 mg/L IBA + 2.0 mg/L Kn when compared to all other concentrations and combinations of phytohormones tested (Table 6.5). Addition of auxin, IBA and cytokinin, Kn together in the medium induced maximum number of shoots/explant compared to nodal buds cultured on MS medium supplemented with Kn alone (Figures 6.4 and 6.6).

Rooting was best achieved on ½ MS medium fortified with IBA (0.1–1.0 mg/L) and IAA (0.1–1.0 mg/L). Rooting was also induced from micro shoots on ½ MS medium alone but the root was single, thin and short in comparison to roots developed in all other hormonal concentrations studied (Table 6.6). The plantlets were washed thoroughly with running tap water and those were transferred to a mixture of soil and compost (1 : 1) for hardening. Later these were shifted to green house (Plate 5d).

The first response exhibited by the nodal segments was the enlargement and breaking of the axillary buds. During the proliferation stage both axillary and adventitious shoots grewout from the original explant. In the event of multiple shoots production, new adventitious buds and shoots developed form the base of the pre-existing ones was observed. These new shoots when cut into

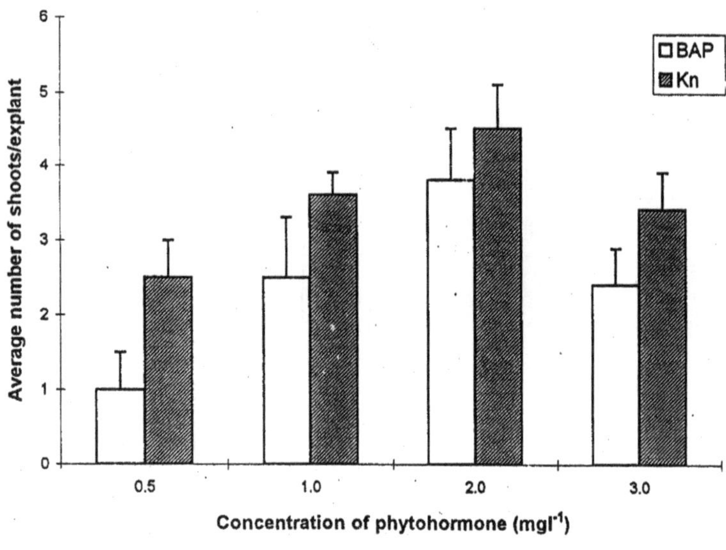

Figure 6.5: Effect of Different Concentrations of BAP/Kn on Multiple Shoot Induction from Nodal Bud Cultures of *S. surattense*

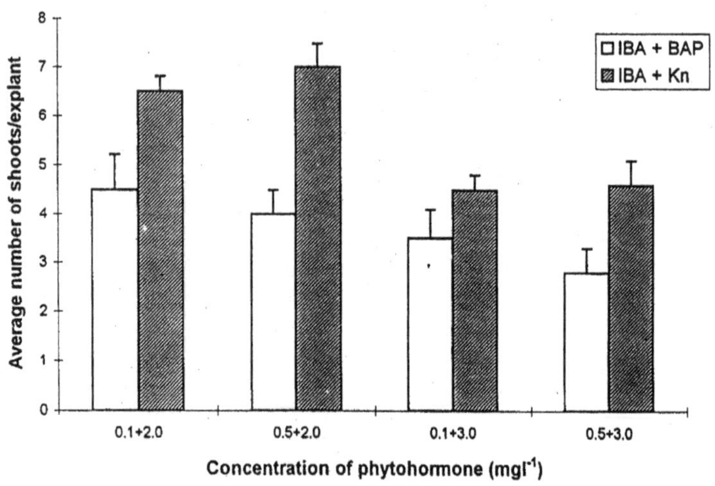

Figure 6.6: Effect of Different Concentrations of IBA+BAP/Kn on Multiple Shoot Induction from Nodal Bud Cultures of *S. surattense*

segments containing one or two nodes and cultured on the same medium, gave rise to new shoots. Thus, from a single explant it was possible to obtain a number of *in vitro* regenerated shoots, within a period of 6 weeks in *S. surattense* (Ramaswamy *et al.*, 2004).

Influence of explanting season on culture establishment was also noted in *Tridax procumbens* (Sahoo and Chand, 1998) as we have observed in *S. surattense*. Similarly, this was shown in other medicinal herbs including *Ocimum* species (Ahuja *et al.*, 1982; Pattnaik and Chand, 1996).

Multiple shoots were induced on MS medium supplemented with various concentrations of cytokinins such as BAP and Kn alone and also in combination with auxins IBA except at the concentration of 0.5 mg/L BAP in *S. surattense*. Similarly, Sudershan *et al.*, (2000) have observed the multiple shoot bud induction from nodal segments of *Zizyphus mauritiana* on MS medium supplemented with BA alone. It was also recorded the same results in *Vanilla planifolia* on MS + BAP alone (Geetha and Shetty, 2000). When BAP and Kn concentration was increased (above 2.0 mg/L), the rate of shoot multiplication and elongation was reduced in the present investigations. Similar results were obtained in *Canavalia virosa* (Kathiravan and Ignacimuthu, 1999), and *Pisonia alba* (Jagadishchandra *et al.*, 1999). Sahoo and Chand (1998) have also found that the percentage bud break and induction declined with the increase (above 2.0 mg/L) in BAP concentration. Our observations are also in conformity with those of Ahuja *et al.* (1982) in *Ocimum gratissimum* and *O.viride* and Pattnaik and Chand (1996) in *O.americanum* and *O. Sanctum.*

At high concentration of cytokinin lateral bud break was suppressed but callus proliferation improved as observed by Shahzad and Siddique (2000) and also reported by Ahuja et al, (1982), Pattnaik and Chand (1996) and Sahoo and Chand (1998).

It has been found that with higher concentration of cytokinins growth of individual shoots was arrested and required an additional GA_3 for *in vitro* elongation, as observed in sandalwood plantlets and in *Canavalia virosa* (Kathiravan and Ignacimuthu, 1999). It was also interesting to note that the increased concentration of GA_3 (above 2.0 mg/L) reduced the shoot proliferation and elongation (Kathiravan and Ignacimuthu, 1999).

The medium containing Kn showed the superiority over BAP in inducing multiple shoots in *S. surattense* as observed by Skoog and Miller (1957) in tobacco. Whereas Sahoo and Chand (1998) observed the effectiveness of BAP over Kn in inducing bud break as well as multiple shoot formation in *Tridax.*

Likewise, Zaman *et al.* (1996) have also studied the clonal propagation of mulberry plants and noted the induction of maximum number of multiple shoots/explant from the nodal buds cultured on BAP compared to other cytokinins such as Kn, 2-ip and Zeatin tested. Even, the highest number of shoots (18.9) per explant was obtained when 50 mg/L tyrosine was added with BA. The addition of tyrosine along with three other cytokinins also improved shoot regeneration efficiency but lesser to that of BA in mulberry.

Direct regeneration of multiple shoots from nodal explants as observed on IBA + BAP/Kn supports the findings of Shahzad and Siddique (2000) in *Ocimum sanctum,* Gill *et al.* (1996) in *Azadirachta indica,* Jagadish Chandra and Shathyanarayana (1997) in *Morus indica* and Varisaimohamed *et al.* (1998) in *Macrotyloma uniflorum.* Bais *et al.* (2000) have also observed the maximum number of shoots on MS medium supplemented with auxin + cytokinin combination in nodal cultures of *Decalepis hamiltonii* as it was found in *S. surattense.* The same synergistic effect was also recorded in *Plumbago indica* inducing maximum number (17) of shoots/nodal explant on MS + IAA (0.1 mg/L) + BA (3.0 mg/L).

Likewise various successful combinations have been reported such as IAA + BAP for *Adhatoda beddomei* (Sudha and Seeni, 1994), *Alpinia galanga* (Anand and Hirahardan, 1997), *Duboisia myoporoides* (Kukrejha and Mathur, 1985) and *Valeriana wallichi* (Mathur *et al.,* 1988), IBA + BAP for *Rheum emodi* (Lal and Ahuja, 1989) and *Gardenia jasminoides* (George *et al.,* 1993), NAA + BAP for *Gomphrena officinalis* (Merci er *et al.,* 1992) and *Rauvolfia serpentina* (Mathur *et al.,* 1987); 2, 4-D + BAP for *Withania somnifera* (Sen and Sharma, 1991) as well as BAP + Kn for *Mentha spp.* (Rech and Pires, 1986), *Kaempfera galanga* (Vincent *et al.,* 1992) and, *Feronia limonia* (Hossain *et al.,* 1994).

Malathy and Pai (1998) have recorded that adenine sulphate was found to be essential for axillary shoot initiation in *Ixora singaporensis.* One reason could be the inefficiency of cytofemin synthetase in the plant that it requires adenine in order to counter the senescence of the lateral bud.

Though several workers have reported the requirement of both auxins and cytokinins for induction of shoots (Das and Mitra, 1990; Roy *et al.*, 1993); Seetharam *et al.*, (2003) have used the cytokinins together *i.e.* BA (1.0 mg/L–4.0 mg/L) + Kn (0.5mg/L–1.0 mg/L) for nodal culture. They observed that BAP/Kn alone failed to eticit multiple shoots from nodal explants. Cytokinins alone such as BAP/ Kn were sufficient to initiate adventitious buds in nodal cultures of *S. surattense* (Ramaswamy *et al.*, 2004). Similary, this stimulatory effect of a singular supplement of cytokinin was reported earlier in other medicinal species including *Curcuma spp.* and *Zingiber officinale* (Balachandran *et al.*, 1990), *Chlorophytum borivillianum* (Purohit *et al.*, 1994), *Piper sp.* (Bhatt *et al.*, 1995), *Ocimum spp.* (Pattnaik and Chand, 1996; Sahoo *et al.*, 1997) and *Tridax procumbens* (Sahoo and Chand, 1998).

Thus, direct multiple shoot production was observed from nodal segments in *S. surattense*. This type of clonal propagation has advantage by producing true-to-type of plants from a single individual in a relatively short time (Figure-6.2). Micropropagation using shoot meristem/nodal bud culture shows the considerable importance for large-scale propagation of *S. surattense*, an important medicinal plant.

Chapter 7

Floral Bud Culture

Floral bud culture *in vitro* offers an unique technique where by the influence of vegetative parts could be eliminated facilitating analysis of the role of nutritional, hormonal and environmental factors in successful morphogenesis (Konar and Kitchlue, 1982). Floral bud culture *in vitro* was reported in a number of species (Dunstan and Short, 1979; Pike and Yas, 1990; Novak and Howel, 1981; Neeta *et al.*, 2000; Rauber Grounewaldt, 1988; Niranjan and Sudarshana, 2000). Efficient plant regeneration was through floral bud culture achieved *in vitro* (Rama Swamy *et al.*, 2006c).

The results on floral bud culture in *S. surattense* cultured on various concentrations and combinations of cytokinins and auxins are shown in Figsures 7.1–7.4.

Within the first week, most of the inoculated flower buds enlarged. After 2 weeks of culture they opened followed by the enlargement of the ovary. Adventitious shoot buds were induced from the ovary region after 3 weeks of culture. Direct shoots were formed from the explants on modified MS medium amended with cytokinins + auxins and also cytokinins alone tested (Rama Swamy *et al.*, 2006c). Maximum number of adventious shoot buds/explant was found at 3.0 mg/l BAP/Kn when it was added alone to the medium, but highest percentage of responding cultures and more

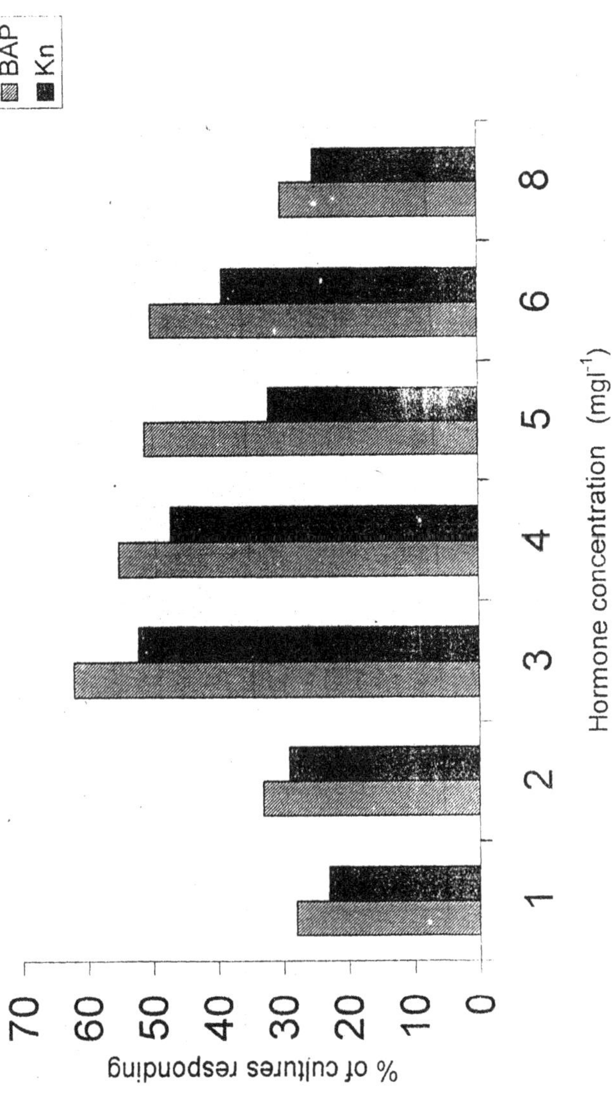

Figure 7.1: Effect of BAP/Kn on Induction of Multiple Shoots from Floral bud Cultures of *S. surattense*

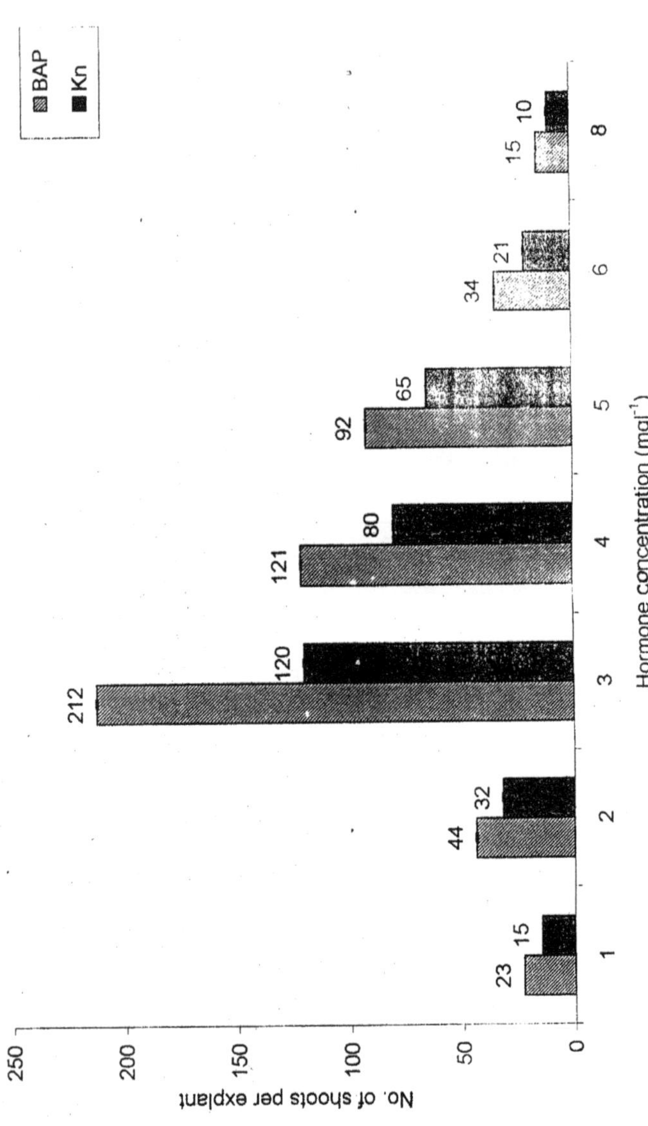

Figure 7.2: Effect of BAP/Kn on Induction of Multiple Shoots from Floral bud Cultures of *S. surattense*

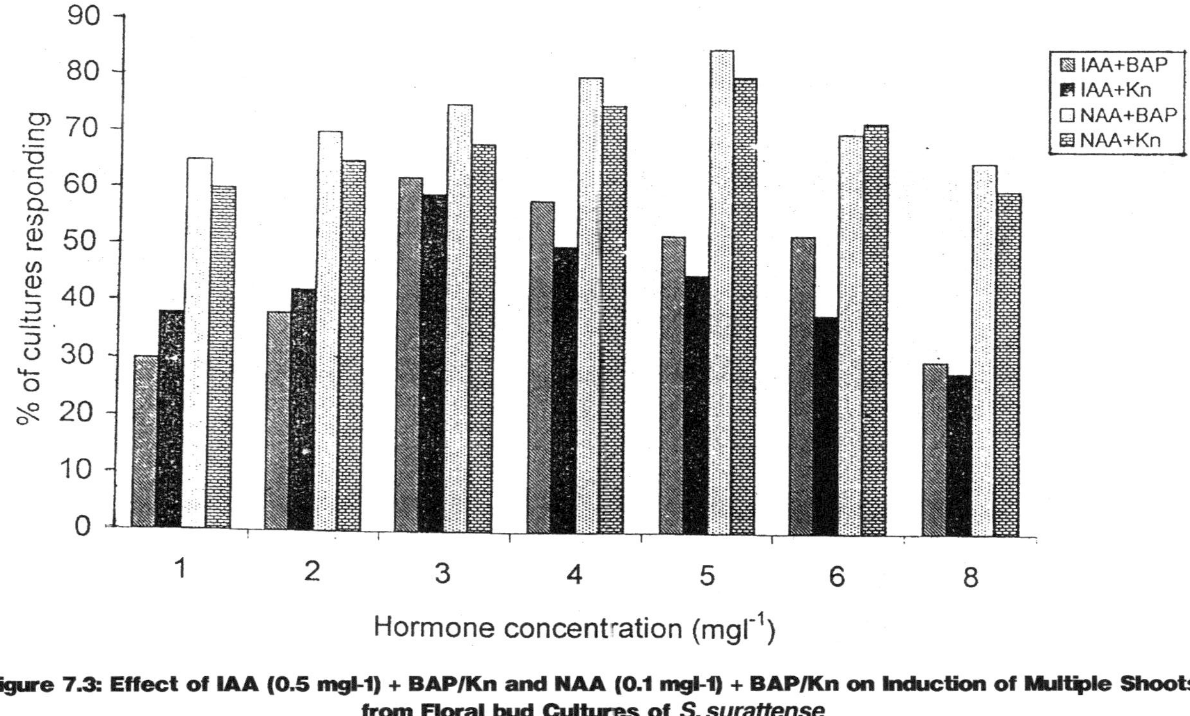

Figure 7.3: Effect of IAA (0.5 mgl-1) + BAP/Kn and NAA (0.1 mgl-1) + BAP/Kn on Induction of Multiple Shoots from Floral bud Cultures of *S. surattense*

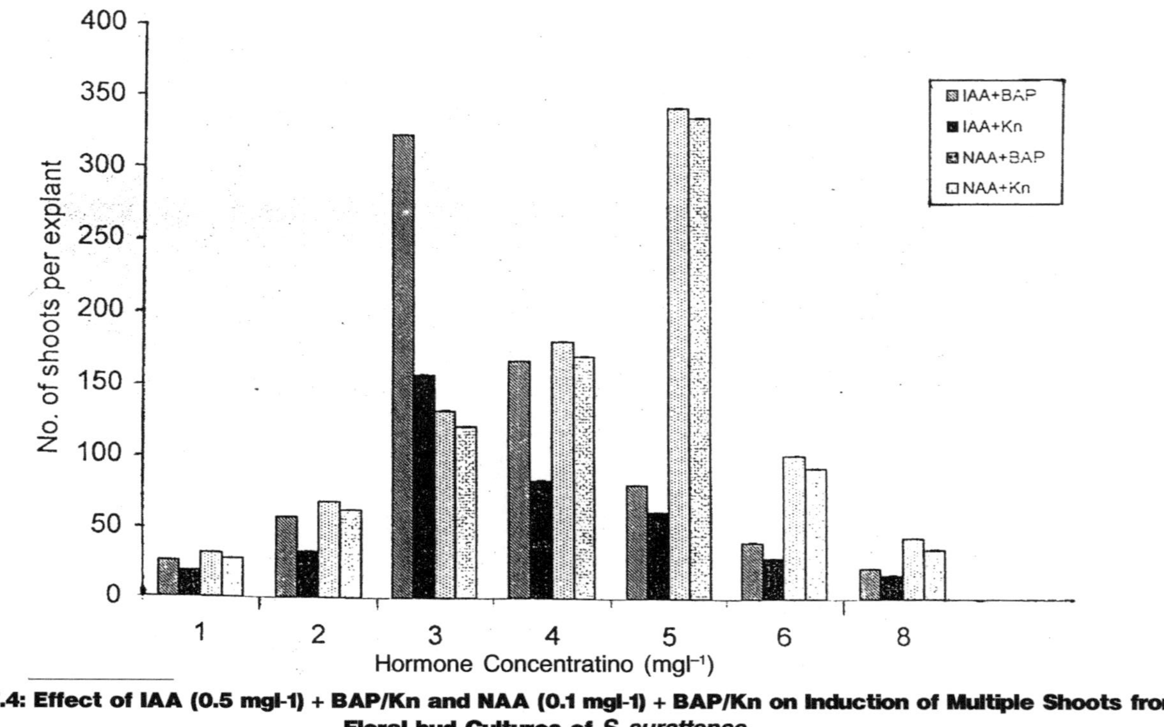

Figure 7.4: Effect of IAA (0.5 mgl-1) + BAP/Kn and NAA (0.1 mgl-1) + BAP/Kn on Induction of Multiple Shoots from Floral bud Cultures of *S. surattense*

number of shoots were recorded at 3.0 mg/l BAP (212 ± 0.12). As the concentration increased the frequency of response and average number of shoots induction have also been increased upto 3.0 mg/L BAP/Kn. Whereas at high concentrations the percentage of responding cultures and shoot bud induction were found to be decreased gradually (Figures 7.1 and 7.2). At 8.0 mg/L BAP/Kn very less number of shoots was developed compared to other concentrations of BAP/Kn alone.

Effect of IAA (0.5 mg/L) along with the various concentrations of BAP/Kn on morphogenesis of flower bud was also tested (Figures 7.3 and 7.4). Modified MS medium containing 0.5 mg/L IAA in combination with BAP/Kn (1.0–8.0 mg/L) showed the enhanced efficiency in inducing the adventitious shoot buds/explant. IAA (0.5 mg/L) in combination with 3.0 mg/L BAP/Kn produced maximum number of shoots (Plate 6a) with highest responding frequency compared to other concentrations of BAP/Kn. As the concentration of BAP/Kn increased in the medium showed the less response and decreased number of shoots gradually from 4.0 mg/L BAP/Kn + 0.5 mg/L IAA combination onwards.

The response of floral buds *in vitro* was observed using modified MS medium containing various concerntrations of cytokinins such as BAP/Kn in combination with 0.1 mg/L NAA. Direct shoot bud proliferation was found in all the concentrations of BAP/Kn with 0.1 mg/L NAA tested (Figures 7.3 and 7.4). More number of adventitious shoots/explant was recorded at 0.1 mg/L NAA + 5.0 mg/L BAP/Kn compared to all other concentrations of cytokinins used (Plate 6b). Low concentration of BAP/Kn induced less number of shoots/explant but gradually the shoot bud induction was found to be increased upto 5.0 mg/L BAP/Kn + 0.1 mg/L NAA and at high concentrations the shoot bud proliferation from floral bud cultures was reduced. High percentage of responding cultures and maximum number (323 ± 0.35) of shoots were observed on MS modified medium supplemented with 0.1 mg/L NAA + 5.0 mg/L Kn compared to all other combinations and concentrations studied (Figures 7.3 and 7.4).

The microshoots were excised and individually transferred to MS medium containing 3 per cent (w/v) sucrose and augmented with 0.5 to 1.5 mg/L IAA for root induction. Root initiation was profuse in the medium containing 1 mg/L IAA.

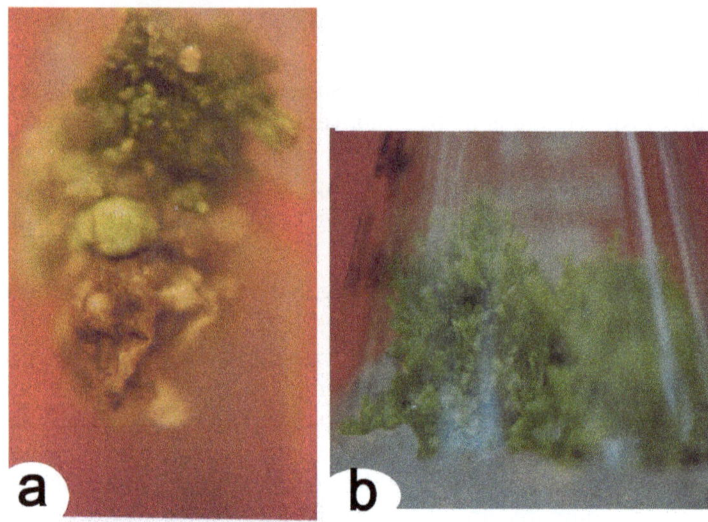

Plate 6a: Multiple Shoots Formation from Floral Bud on MS Modified Medium Supplemented with 0.5 mg/L IAA + 3.0 mg/L Kn in *S. surattense.* (Note the remains of petals).

Plate 6b: Induction of Thousands of Shoot Buds from a Single Floral Bud After Subculture on 0.1 mg/L NAA + 5.0 mg/L Kn in a Conical Flask

Plate 6c: Hardened Plant Developed from Floral Bud Culture

The plantlets were transferred to plastic pots containing sterile soil: compost mixture (1 : 1) and kept for 15 days in culture room at 25 + 1⁰C. Subsequently these were transferred to earthernware pots containing mixture of soil and compost (1 : 1) and shifted to green house (Plate 6c).

Similar direct shoot regeneration from floral buds was also reported in *Allium cepa* (Dunstan and Short, 1979; Pike and Yas, 1990) and *A. porrum* (Novak and Havel, 1981; Rauber and Groune Waldt, 1988) as it was observed in *S. surattense*. Ashalatha and Seo (1993) have observed the induction of more shoots on 4.5 µM 2,4-D + 22.2 µM BA medium. Niranjan and Sudarshana (2000) have observed *in vitro* plant regeneration in *Nymphoides* from the callus cultured on MS + 2.0 mg/L BA + 1.0 mg/L 2,4-D. Direct plantlet regeneration was obtained in *Curcuma longa* a medicinally important plant on MS medium supplemented with both cytokinins + auxins such as BA in combination with 2,4-D or NAA and TDZ or in combination with IAA as in *S. surattense*. They have found highest number of shoots/explant (8.9) on TDZ (2.0 mg/L) + IAA (1.0 mg/L). Since floral buds are modified vegetative buds, it is quite likely that immature floral buds have the capacity to revert back to vegetative buds under appropriate *in vitro* conditions (George and Sherrington, 1984).

Adventitious shoot regeneration in *S. surattense* has been reported with 213 and 148 shoot buds per explant from leaf and shoot tip cultures respectively (Pawar *et al.*, 2002) and 25.8 shoots from leaf and 23.6 shoots from nodal explants (Seetharam *et al.*, 2003). Whereas in the present investigation the number of shoots developed directly from the explant 323. When the same was subcultured in a conical flask containing the fresh medium induced thousands of shoots per explant.

Thus, plant regeneration *in vitro* from floral bud cultures was obtained in medicinally important herb *S. surattense*, which arise directly from the tissues of the explant can provide a method for micropropagation. The present study shows that floral bud culture is amenable to high frequency regeneration and opens up the possibility of using this pant in genetic manipulation for introduction of genes conferring bacterial and fungal resistance using particle gun bombardment (Rama Swamy *et al.*, 2006c).

Chapter 8

Androgenic Haploidy

One of the very popular methods for production of haploids is through culturing anther on artificial culture medium. This leads to the growth of microspores into sporophytes. After the initial reports of successful regeneration of haploids from culture in *Datura innoxia*(Guha and Maheshwari,1964, 1966, 1967), androgenic haploids have been obtained in more than 150 species belonging to 23 families of Angiosperms (Maheshwari *et al.*, 1980). Haploid plants derived from anther/pollen culture can greatly reduce the time taken to obtain a homozygous line after their chromosomes doubling (Morrison and Evans,1988). This technique plays an important role in plant breeding program.

Several reports have shown the feasibility of production of androgenic lines from anther culture in various species (Isouard *et al.*, 1979; Dumas de Vaulx and Chambonnet, 1982; Rotino *et al.*, 1987).

Haploids derived from anther cultures have considerable potential for plant breeding because of the time saved by the reduction of the classical selection cycle period and the genetic value of isogemic lines or homozygous diploids. The double haploid plants resulting from anther culture are useful for breeding in improving the efficiency with which superior genotypes can be identified

(Knapp, 1991; Mitchell *et al.*, 1992; Bjornastad *et al.*, 1993). Also the presence of a single set of chromosomes allow the detection of mutations controlled by recessive genes. Application of the technique has been intensified and extended to studies of haploid induction in crop plants notably Solanaceous species such as tobacco, tomato, tuberous Solanums and *Capsicum* etc. Rotino (1996) conducted the experiments to obtain double haploid parents in cultivated *Solanum melongena*. Prasad *et al.* (1998) have reported the androgenic stable somaclonal variant (tetraploid) of *S. surattense*. Androgenic haploidy has been achieved through anther culture in *S. surattense* (Ramaswamy *et al.*, Pers. comm.).

The anthers which were excised during August–September and cultured on AC medium supplemented with different concentrations of auxins and cytokinins showed better response in inducing callus or shoot buds compared to anthers collected during other periods.

The results on morphogenic response of anthers cultured on AC medium containing various concentrations of cytokinins in combination with auxins such as 0.1 mg/L NAA + BAP/Kn, 1 mg/L NAA + BAP/Kn and 0.1 mg/L IAA + BAP/Kn, 1 mg/L IAA + BAP/Kn are shown in Tables 8.1-8.4. Anthers initially showed slight swelling on various media combinations and concentrations and during the second week of incubation they started callusing. Percentage of frequency of callusing was gradually increased from 1.0 mg/L to 3.5 mg/L cytokinins tested in combination with NAA/IAA. High percentage of frequency of callusing was noted at 3.5 mg/L BAP/Kn with 0.1 mg/L IAA/NAA or 1.0 mg/L IAA/NAA. With in 2-3 weeks after culture, callus emerged after bursting of mid region of anther lobes and even sometimes from the lower region of anther lobes or occasionally along the entire microsporangial surface (Plate 7a).

At high concentrations of cytokinins showed the reduced percentage of callus production. After five to six weeks of anther culture more quantity of callus induction was observed at 0.1 mg/L IAA/NAA in combination with 3.5 mg/L BAP/Kn and at 1 mg/L IAA/NAA + 3.5 mg/L BAP/Kn. Maximum callus growth, highest frequency of responding cultures with embryogenic callus were found after 6 weeks of culture at 1.0 mg/L NAA in combination with 3.5 mg/L Kn compared to all other combinations and concentrations of auxins and cytokinins tested.

TABLE 8.1
Morphogenic Response of Anther Culture of *S. surattense* on AC Medium Supplemented with NAA + BAP

Growth Regulators (mg/L)			% of Cultures Responding	Morphogenic Response
NAA	+	BAP		
0.1	+	1.0	23	Brown callus
0.1	+	1.5	41	Callus
0.1	+	2.0	64	Callus
0.1	+	2.5	70	Nodular callus
0.1	+	3.0	90	Nodular callus
0.1	+	3.5	95	Nodular callus
0.1	+	4.0	90	White callus
0.1	+	4.5	83	White callus
0.1	+	5.0	60	White callus
1.0	+	1.0	32	Callus
1.0	+	1.5	53	Callus
1.0	+	2.0	65	Nodular callus
1.0	+	2.5	80	Nodular callus
1.0	+	3.0	83	Green nodular callus
1.0	+	3.5	90	Green nodular callus
1.0	+	4.0	85	Green nodular callus
1.0	+	4.5	72	Green callus
1.0	+	5.0	60	Green callus

Plant Regeneration

Calli derived from anther cultures grown on AC medium containing 1.0 mg/L NAA + 3.5 mg/L Kn were transferred to fresh AC medium augmented with 0.1 mg/l TDZ in combination of various concentrations of BAP/Kn (Table 8.5). Calli cultured on 0.1 mg/L TDZ with 1.0–4.0 mg/L BAP showed the callus induction whereas at 5.0 mg/L BAP induced the shoot bud production by developing 2-3 green + 2 albino shoots. Anther calli when cultured on AC medium supplemented with 0.1 mg/L TDZ and 1.0 to 6.0 mg/L Kn showed the plant regeneration in all the concentrations studied. Maximum (12-16) shoot bud proliferation/explant was observed at

Plate 7a: Induction of Embryogenic Callus on ALM + 1.0 mg/L NAA + 3.5 mg/L Kn in *S. surattense*

Plate 7b: Multiple Shoots Formation from Androgenic Callus on ACM + 0.1 mg/L TDZ + 2.0 mg/L Kn (Note the albino shoots)

Plate 7c: Androgenic Haploid Plant in a Plastic Pot After Hardening

0.1 mg/L TDZ + 5.0 mg/L Kn with high frequency of regeneration efficiency (Plate 7b) compared to all other concentrations of Kn and BAP too. At high concentration of Kn, albino plants were formed.

TABLE 8.2
Morphogenic Response of Anther Culture of *S. surattense* on AC Medium Supplemented with NAA + Kn

Growth Regulators (mg/L)			% of Cultures Responding	Morphogenic Response
NAA	+	Kn		
0.1	+	1.0	35	Callus
0.1	+	1.5	43	Callus
0.1	+	2.0	60	Callus
0.1	+	2.5	75	Callus
0.1	+	3.0	83	Nodular callus
0.1	+	3.5	95	Nodular callus
0.1	+	4.0	82	Nodular callus
0.1	+	4.5	75	Callus
0.1	+	5.0	60	Callus
1.0	+	1.0	54	Callus
1.0	+	1.5	68	Callus
1.0	+	2.0	72	Nodular callus
1.0	+	2.5	75	White friable callus
1.0	+	3.0	92	Green embryogenic callus
1.0	+	3.5	97	Green embryogenic callus
1.0	+	4.0	90	Green nodular callus
1.0	+	4.5	82	Callus
1.0	+	5.0	73	Callus

The callus which was compact did not induce any plantlet regeneration whereas white/green-embryogenic/friable calli were found be effective in inducing the shoot buds in the present studies.

It is clearly indicated that TDZ induced the shoot bud proliferation from calli derived from anthers and also showed more pronouncement of regeneration efficiency when combined with Kn rather than BAP in *S. surattense*.

TABLE 8.3
Morphogenic Response of Anther Culture of *S. surattense* on AC Medium Supplemented with IAA + BAP

Growth Regulators (mg/L)			% of Cultures Responding	Morphogenic Response
IAA	+	BAP		
0.1	+	1.0	23	Callus
0.1	+	1.5	30	Callus
0.1	+	2.0	53	Callus
0.1	+	2.5	64	Callus
0.1	+	3.0	70	Green callus
0.1	+	3.5	75	Green callus
0.1	+	4.0	60	Callus
0.1	+	4.5	53	Callus
0.1	+	5.0	41	Callus
1.0	+	1.0	53	Callus
1.0	+	1.5	61	Callus
1.0	+	2.0	65	Callus
1.0	+	2.5	70	Callus
1.0	+	3.0	80	Green callus
1.0	+	3.5	89	Green callus
1.0	+	4.0	60	Callus
1.0	+	4.5	58	Callus
1.0	+	5.0	53	Callus

In vitro Rooting of Anther Culture Derived Plantlets

Micro-shoots developed from anther calli cultures were isolated and transferred to rooting medium containing ½ strength MS medium supplemented with different concentrations of IAA (0.1–0.5 mg/l). Among these concentrations 0.1 mg/l IAA proved to be better by inducing more number of roots (10-12) after 4 weeks of culture. *In vitro* rooted plants were washed gently under running tap water to remove the excess medium. These were transferred to plastic pots containing sterile soil mix *i.e.*, soil : compost (2 : 1) and hardened in the culture room under 80–90 per cent humidity (Plate 7c). Later these plants were transferred to the green house. These androgenic plants were found to be haploids.

TABLE 8.4
Morphogenic Response of Anther Culture of *S. surattense* on AC Medium Supplemented with IAA + Kn

Growth Regulators (mg/L)			% of Cultures Responding	Morphogenic Response
IAA	+	Kn		
0.1	+	1.0	19	Callus
0.1	+	1.5	21	Callus
0.1	+	2.0	42	Callus
0.1	+	2.5	51	Callus
0.1	+	3.0	56	Green callus
0.1	+	3.5	65	Green callus
0.1	+	4.0	60	Callus
0.1	+	4.5	54	Callus
0.1	+	5.0	46	Callus
1.0	+	1.0	41	Callus
1.0	+	1.5	55	Callus
1.0	+	2.0	60	Callus
1.0	+	2.5	62	Callus
1.0	+	3.0	67	Green callus
1.0	+	3.5	70	Green callus
1.0	+	4.0	61	Callus
1.0	+	4.5	53	Callus
1.0	+	5.0	48	Callus

The anthers cultured on AC medium fortified with various concentrations of cytokinins (BAP/Kn) in combination with auxins (NAA/IAA) showed the callus induction without any pre-treatment of anther in *S. surattense*. Highest frequency percentage of calli growth was found at 1.0 mg/L NAA + 3.5 mg/L Kn in comparison to all other combinations and concentrations tested. Similarly, Mandal and Gupta (1997) have noted the rate of callusing was highest in rice hybrid anthers cultured on 2 mg/L NAA + 0.5 mg/L Kn. For induction of callus in anther cultures of Niger, auxin (2 mg/L 2,4-D) in combination with cytokinin (0.3mg/L Kn) was used (Sarvesh *et al.*, 1994). Virendra *et al.* (1993) have also found the essentiality of auxin + cytokinin combination in *Azadirachta indica* a medicinal tree. According to their observations a higher percentage of cultures produced callus on Nitsch medium containing IAA + BA. Maximum percentage of callus induction from anther cultures was reported in *Manihot esculenta* by using 2 mg/L each of NAA + 2,4-D and Kn (Abraham *et al.*, 1995). Martha *et al.* (1991) proved that the medium containing 1 mg/L NAA + 0.1 mg/L BA was the best among the tested combinations and concentrations of growth regulators. Prasad *et al.* (1998) have also observed the same findings in *S. surattense* when anthers cultured on Nitsch and Nitsch medium as it was found in *S. surattense* that the requirement of both auxins and cytokinins. The same phenomenon was also observed in *S. melongena* (Dumas de Vaulx and Chambonnet, 1982).

TABLE 8.5
Effect of TDZ in Combination with BAP/Kn on Induction of Multiple Shoots from Anther Callus Cultures of *S. surattense*

Hormone Conc. (mg/L)			% of Cultures Responding	Morphogensis and Regenerated Shoots/ Explant
IAA	+	Kn		
TDZ	+	**BAP**		
0.1	+	1.0	65	Callus
0.1	+	2.0	68	Callus
0.1	+	3.0	75	Callus
0.1	+	4.0	80	Callus
0.1	+	5.0	85	2-3 Green plants + 2 albino plants
0.1	+	6.0	73	Green callus
TDZ	+	**Kn**		
0.1	+	1.0	60	One green plant
0.1	+	2.0	75	One green plant + albino plant
0.1	+	3.0	80	2-4 Green plants
0.1	+	4.0	85	6-8 Green plants + 4 albino plants
0.1	+	5.0	90	12-16 Green plants + 2 albino plants
0.1	+	6.0	70	7 albino plants

The embryogenic calli induced on AC medium augmented with 1.0 mg/L NAA + 3.5 mg/L Kn when transferred to media containing

0.1 mg/L TDZ + BAP/Kn (1.0 –6.0 mg/L) showed multiple shoot bud proliferation at 3.0-5.0 mg/L Kn. Kn had showed the superiority in inducing more number of shoots compared to BAP in all the concentrations tested in combination with TDZ (0.1 mg/L) in anther cultures of *S. surattense*. Similarly, Prasad *et al.* (1998) have also found the differentiation of progressively more shoot buds with its increasing concentrations from 0.25 to 3.0 mg/L in anther cultures of *S. surattense*. Sarvesh *et al.* (1994) obtained the shoot regeneration on MS medium fortified with high level (1.0 mg/L) of cytokinin (BAP) and less concentrations of auxin (0.1 mg/L NAA) in niger. Stoehr and Zsuffa (1990) observed the shoot bud induction from anther calli on high levels of auxin (2,4-D) in combination with low level of cytokinin (Kn) in *Populus maximowiczii*. Whereas in *Scabiosa columbaria*, it is interesting to note that stepwise removal of growth regulators and simultaneous lowering of sucrose from the nutrient medium resulted in initiation of embryogenesis or shoot organogenesis and allowed plant regeneration from anther calli cultures (Romeij and Van Lammeren, 1999).

Virendra *et al.* (1993) have found the multiple shoots induction in the anther derived callus when subcultured on MS medium supplemented with 4.4 mm BA + 0.53 mm NAA in *Azadirachta indica*. Thus Cai *et al.* (1992) indicated that plant regeneration dependent on the type of concentration of growth regulators used for callus induction. According to Chee (1992) and Hu *et al.* (1993) plants differ in their growth regulators requirements for inducing anther callus/regeneration.

However, the success in anther culture technology seems to be highly dependent on pretreatment period, genotype, environment (temperature, light intensity, photoperiod), culture conditions, physiology/growth of anther-donor plant and their interactions (Ekiz and Konzak, 1991, 1993, 1994; Beerasft and Taylor, 1992; Zhou and Konzak, 1992; Moieni and Sarrafi, 1995; Hasan and Konzak, 1997) in addition to the different types of media. Sarvesh *et al.* (1994) have noted the influence of both genotype and treatment affected the induction of embryogenic and non-embryogenic callus from anthers of niger. Similar genotype dependence was found in androgenic response of pepper (Gomez and Chambonnet, 1992; Qin and Rotino, 1993; Mitgko *et al.*, 1995). Rotino (1996) also found the effect of stage of anther, genotype influence and culture conditions in androgenic haploid production of *Solanum melongena.* Haploid plants also successfully been generated from microspores of eggplant (Gu, 1979; Miyoshi, 1996). The best results were obtained after a pretreatment period of 3 days at 35^{0}C, in a medium lacking sucrose (Miyoshi, 1996). Sucrose starvation was reported to suppress gametophytic development and DNA synthesis of cultured microspores. A short period of high temperature during the initial culture was found to inhibit the normal development of the microspores, and then stimulated androgenesis.

The effect of donor plant environment was first emphasized by Kristiansen and Andersen (1993). Similarly, it was also observed in *S. surattense* that the anthers did not respond well when they were collected during October to July (Ramaswamy *et al.,* Pers. Comm.).

Thus, it is evident that the ability to induce callus and regenerate plants from anthers of *S. surattense* through callus mediated shoot organogenesis. The protocol allows efficient regeneration, to be used for plant breeding, fundamental research or cultivation and production purposes.

Chapter 9

Salt and Drought Tolerance

Abiotic stresses such as higher salt levels, low water availability, excess water and high and low temperatures adversely affect growth and yield of crop plants (Skriver and Mundy, 1990; Kavi Kishor *et al.*, 1995; Grover *et al.*, 1998a,b). In order to improve the quality and yield of the crop, there is a great deal of urgency in improving the performance of crop against their abiotic stress factors. Stress effects on plant cells and tissues were investigated and the stress inducing compounds used in different experiments were either ionic and penetrating (*e.g.* NaCl), non-ionic and penetrating (*e.g.* mannitol; sorbitol) or non-ionic and non-penetrating (eg. polyethylene glycol, Gangopadhyay *et al.*, 1997a). Results of such experiments, however, have shown the various stress effects in the plant tissues, such as i) retardation of growth (Rains, 1989; Thomas *et al.*, 1992), ii) acquisition of the ability of adapt to stressfull environments (Binzel *et al.*, 1985; Handa *et al.*, 1986) and iii) accumulation of proline or osmolytes at high level (Chandler and Thorpe, 1987; Paek *et al.*, 1988; Jain *et al.*, 1991a,b; Thomas *et al.*, 1992; Verbruggen *et al.*, 1993; Gangopadhyay *et al.*, 1997a, b; Shah *et al.*, 2002). The osmolytes, including fructans, trehalose, mannitol, ononitol, myo-inositol, glycine betaine and polyamines accumulate in significant amounts which might be helpful in maintaining osmolyte potential, ionic balance, membrane integrity, Oxygen free radical levels and chromation protection under the stress conditions (Bohnert and Jensen, 1996; Rajam *et al.*, 1998; Bohnert and Shen, 1999).

Genetic modification of crop plants to improve their salt tolerance is a possible way of increasing production especially for regions of the semi-arid tropics. Success for obtaining tissue culture variant lines that have been attained in some crop plants (Croughan *et al.*, 1981; Van Swaaij *et al.*, 1986; Mc Coy, 1987; Mc Hugen, 1987; Jain *et al.*, 1990; 1991 a,b; Tal, 1994).

In vitro screening for salt and drought tolerance would seem to be particularly suitable since NaCl/KCl/mannitol can be easily incorporated into the medium. However, despite the considerable effort devoted to *in vitro* screening for salt tolerance, plant regeneration from selected tolerant cell lines has proved to be a major barrier. One approach that overcomes the problem of regeneration of selected cell lines is to carryout the *in vitro* screening with differentiating cultures (Ibrahim *et al.*, 1992).

The main cause of difficulty in producing salt–and drought tolerant regenerants may be physiological adaptation of *in vitro* cultures instead of mutation coupled with lack of fertility. However, recent results suggest that difficulties may over come using explants with high regeneration potential and by applying short-term selection (Dorion *et al.*, 1999).

Recently, Rama Swamy *et al* (2006d) have screened for salt (NaCl, KCl) and mannitol tolerance in *S. surattense* using an *in vitro* selection.

The leaf explants when cultured on RM (regeneration medium) containing NaCl (0.1 per cent to 1.0 per cent) showed the gradual decrease in the production of plantlets/explant which were under NaCl stress (Plate 8a). At 1.0 per cent of NaCl the total regeneration was inhibited. The percentage of responding cultures was also decreased from 0.1 per cent NaCl to 0.75 per cent NaCl where 90 per cent response was inhibited. At this percentage of NaCl 3.2±0.31 salt-resistant/tolerant shoot buds were developed (Table 9.1).

Under KCl stress too, the leaf explants showed similar decrease in percentage of response and as well as number of shoots

Plate 8a: Comparison of NaCl Tolerant Shoots Induction on 0.1 per cent, 0.2 per cent, 0.3 per cent, 0.4 per cent and 0.5 per cent NaCl Respectively in *S. surattense*

Plate 8b: Comparison of Mannitol Tolerant Shoots Formation on 0.1 per cent, 0.2 per cent, 0.3 per cent, 0.4 per cent and 0.5 per cent Mannitol Respectively

(Note: Rhizogenesis at 0.4 per cent and 0.5 per cent).

Plate 8c: Acclimatized *in vitro* Regenerated Salt Tolerant Plantlet

regeneration/explant. KCl showed more stress compared to NaCl, because the regeneration of shoots/explant was found to be less in number in all the concentrations studied. At 0.75 per cent KCl, only 1.4 + 0.22 shoots were observed with 12 per cent of response (Table 9.1).

TABLE 9.1
Effect of Different Concentrations of NaCl and KCl on Shoot Formation in Leaf Cultures on Regeneration Medium (RM) of *S. surattense*

Conc. of Salt (%)	% of Cultures Responding	Average Number of Shoots/Explant (SE)*
Nacl		
0.1	75	16.0 ± 0.23
0.2	80	14.0 ± 0.35
0.3	60	12.0 ± 0.34
0.4	55	7.5 ± 0.27
0.5	50	5.4 ± 0.43
0.75	10	3.2 ± 0.31
1.0	—	—
KCl		
0.1	75	12.0 ± 0.34
0.2	65	10.5 ± 0.32
0.3	60	8.4 ± 0.34
0.4	55	5.3 ± 0.23
0.5	52	3.2 ± 0.35
0.75	12	1.4 ± 0.22
1.0	—	—

* Mean; ±: Standard Error.

When the leaf explants were cultured on RM containing various concentrations of mannitol (0.1 per cent to 1.25 per cent) the percentage of response was decreased gradually as the mannitol concentration was increased (Table 9.2). At high concentration of mannitol (1.25 per cent), the drought stress was found to be higher, hence the plant regeneration was inhibited. It was also interesting to note that two types of culture response was observed. In some of the cultures, callus and shoots and in other callus + shoots + roots were induced (Plate 8b). The number of shoots/explant was decreased as the concentration of mannitol increased and at 1.0 per cent mannitol 6 ± 0.21 drought tolerant shoots were developed per explant in responded cultures.

TABLE 9.2
Morphogenetic Response of Leaf Explants of *S. surattense* on Regeneration Medium (RM) Containing Different Concentrations of Mannitol

Mannitol Conc. (%)	Percentage of Cultures Responding	% of Cultures with Callus + shoots	% of Cultures with Callus + Shoots + Roots	No. of Shoots/ Explant (S.E.)*
0.1	80	45	35	18 ± 0.33
0.2	75	40	35	16 ± 0.26
0.3	67	35	32	13 ± 0.10
0.4	55	25	30	12 ± 0.18
0.5	45	18	27	10 ± 0.34
1	10	2	8	06 ± 0.21
1.25	—	—	—	

* Mean; ±: Standard Error.

Screening of NaCl/KCl/Mannitol Tolerance

For confirmation and stability of the NaCl/KCl/mannitol tolerance, the leaf explants (50) were excised from the *in vitro* selected (stressful) shoots and cultured on RM containing 0.75 per cent NaCl/ KCl or 1.0 per cent mannitol. The percentage of response was found to be higher (>90 per cent) and also the number of shoot buds proliferated was also observed to be more (NaCl = 12 ± 0.57; KCl = 10 ± 0.31; mannitol 15 ± 0.68) when compared to initial experiments.

For proliferation the selected microshoots and unselected microshoots were cultured on RM containing 0.75 per cent NaCl/ KCl or 1.0 per cent mannitol but with cytokinin *i.e.*, BAP. Proliferation was observed in the selected shoots, whereas it was not found in the unselected shoots and these did not survive after 3 weeks of culture. From these experiments it is clear that the selected showed stable tolerance to the salt or drought stress.

In vitro Rooting from NaCl/KCl/Mannitol Tolerant Shoots

The microshoots developed from selective medium and unselective medium were cultured on rooting medium containing 0.75 per cent NaCl/KCl or 1.0 per cent mannitol (Table 9.3). The selected shoots showed 100 per cent survival but the unselected shoots did not survive. Not significant number of roots was developed from selected shoots compared to unselected shoots in all the treatments but more roots were developed in mannitol compared to NaCl/KCl.

TABLE 9.3
Rooting Performance of Microshoots Derived from Selected (Stressful) and Unselected (Stress Free) Regenerants Cultured on MS Medium + IAA + NaCl/KCl/Mannitol

Treatment NaCl/KCl/ Mannitol	Selected		Unselected	
	% of Survival	No. of Roots/ Shoot (S.E)*	% of Survival	No. of Roots/ Shoot (S.E)*
0	100	16 ± 0.75	100	18 ± 1.0
NaCl (0.75%)	100	08 ± 1.5	—	—
KCl (0.75%)	100	06 ± 0.8	—	—
Mannitor (1%)	100	11 ± 1.0	—	—

* Mean; ±: Standard Error.

These salt and drought tolerant plantlets were hardened and established in green house (Plate 8c). The percentage of survival of these tolerant plants was 30 per cent–40 per cent (NaCl/KCl) and 50 per cent (mannitol) (Ramaswamy *et al*, 2006d).

At high concentration of NaCl/KCl/mannitol very less number of shoots/explant were produced which were tolerant. These regenerated shoots were able to grow at high concentration level of salt showing the stability. These results showed that the selected character was sufficiently stable for it to be transmitted (Rama Swamy *et al.*, 2006d). Similarly the salt tolerant plants were developed in eggplant, tomato (Cano *et al.*, 1998), *Coleus blumei* (Ibrahim *et al.*, 1992), *Pennisetum purpureum* (Bajaj and Gupta, 1987) and *Oryza sativa* (Binch *et al.*, 1992; Mandal *et al.*, 2000). At high concentration of salt levels less number of shoots were able to regenerate in *S. surattense*. As the concentration of the salt was increased there was a gradual increase in the stress conditions (NaCl/KCl/mannitol), hence less number of salt/drought tolerant regenerants could survive compared to unselected (stress free) ones. The regeneration ability in a stressful environment imposed due to NaCl/KCl/mannitol was not identical. More number of tolerant shoots were developed at high concentration of mannitol (1 per cent) than that of low concentration of NaCl/KCl (0.75 per cent) whereas at an equal concentration of NaCl/KCl the morphogenesis was inhibited almost (Rama Swamy *et al*, 2006d). Mannitol is a non-ionic osmoticum whereas NaCl/KCl is ionic in nature. The results revealed that this was due to the effect of an ionic osmoticum than a non-ionic one. Similar differences were also noted in NaCl and mannitol adapted callus lines of *Brassica juncea* (Gangopadhya *et al.*, 1997b). Equimolar concentration of sorbitol a compound similar to mannitol, was reported to be less detrimental to tissue viability than NaCl (Eberhardt and Wegmann, 1989). Gangopadhyay *et al.* (1997 a,b) have noted more tissue injury due to NaCl shock than mannitol and PEG shocks indicated that NaCl stress had both ionic and osmotic components (Filho and Sodek, 1988) and due to the added factor of ion-toxicity it was more detrimental than non-ionic mannitol or PEG stress.

During stress conditions (salt/drought), many physiological and biochemical changes take place. These activities may help the plant to survive under the stress conditions. There were many reports on the physiological and biochemical changes occur when plant tissue under the stress enviornment. It is reported that if stress is imposed gradually then plants are able to undergo metabolic changes that enable them to tolerant at an even more intense stress event (Leone *et al.*, 1994).

Subhashini and Reddy (1990) noted the increased enzyme activity (peroxidase, polyphenoloxydase, acid phosphatase and glutamate dehydrogenase) under various concentrations of salt in rice cultivars. All these enzymes activity was increased linearly following NaCl or sea water stress. Gangopadhyay *et al.* (1997a,b) have observed the growth rate of the adapted callus lines of *Brassica* in NaCl–or mannitol containing media was steady and sustainable but it was lower than the growth of the control callus grown on stress-free medium. The same effect was also recorded in *S. surattense* regarding the regeneration capability of stress tolerant plants in stress-ful and stree-free media (Rama Swamy *et al*, 2006d). Similar observations were also made by many researchers (Greenway and Munns, 1980; Rains, 1989; Cushman *et al.*, 1990; Thomas *et al.*, 1992). According to Cushman *et al.* (1990) this reduced effect under stress was due to certain amount of the total energy available for tissue metabolism was channeled to resist the stress.

Recent studies revealed that salt/drought-tolerant (stress tolerant) somaclonal lines produced more proline inorder to resist the stress environment. Proline over producing somaclones have been reported to be more salt-tolerant than the parent cultivar in *Brassica juncea* itself (Jain *et al.*, 1991a,b). Osmotic stress-induced accumulation of proline has been reported in many species (Watad *et al.*, 1983; Chandler and Thorpe, 1987; Jain *et al.*, 1991a; Verbruggen *et al.*, 1993; Gangopadhyaya *et al.*, 1997 a,b). Growth, viability and proline content of adapted and unadapted callus of *Nicotiana tabacum* var. Jaysri affected due to osmotic stresses and particularly stress-shock treated with different osmotica like NaCl, mannitol and polyethylene glycol were studied and evaluated the physiological differences of stress effects (Gangopadhyay *et al.*, 1997b).

Gangopadhyay *et al.* (1997a, b) have found that the tissues which were under salt/drought stress (NaCl/Mannitol) showed the difference in the accumulation of more free proline for osmoprotection to adapt to an environment with an ionic stress inducing agent than a non-ionic ones. The difference between the effect of an ionic and non-ionic osmoticum on proline metabolism was reported in tobacco callus (Eberhardt and Wegmann, 1989). Proline is accumulated in significant amounts in a variety of plants subjected to salt/drought stresses. Exogenous application of proline protects plant tissues from stress conditions. Increase in proline level is considered to help the cells in osmoprotection as well as in regulating their redox potential, scavenging hydroxy radicals, and in protection against denaturation of various macromolecules (Dubey, 1997).

Higher salt tolerance has been reported in wild tomato species, than in cultivated tomato both at the whole plant level and in callus culture (Perez-Alfocea *et al.*, 1994). On the basis of reduction of both leaf number and shoot fresh weight with increasing salinity, the salt-tolerance of *Lycopersicon esculentum* was lower than that of *L. pennelli* (Cano *et al.*, 1998).

Several genes have been identified which code for enzymes that may be involved in the accumulation of osmolytes during osmotic stress; these include enzymes in the biosynthetic pathway of proline (Delauney and Verma, 1990; Hu *et al.*, 1992), glycine betaine (Weretilnyk and Hanson, 1990), and Polyols (Bartels *et al.*, 1992; Vernon and Bohnert, 1992). The plant biosynthetic pathway for proline contains two genes, one a bifunctional enzyme, pyrroline-5-carboxylate synthetase (Hu *et al.*, 1992) and the other pyrroline-5-carboxylate reductase (Delauney and Verma, 1990). Both of these genes are induced by salt stress, indicating that they may play a role in osmotic adjustment. Induction of salt/drought stress tolerance through genetic engineering technique is emerging area of research. In recent years salt/drought stress tolerant transgenic plants were developed using different types of gene manipulation techniques in various crop species (see Grover *et al.*, 1998 a, b; 2001).

Thus, it is clear from the foregoing discussion that the plant response to stress conditions consists of several events like stress perception, transcriptional activation of stress genes, synthesis and accumulation of stress proteins resulting finally in biochemical, cellular and physiological manifestations. The present protocol is useful for propagation of the species in drought prone areas because of its importance in ayurvedic medicine (Rama Swamy *et al.*, 2006d).

Chapter 10

Antibiotic Resistance

In higher plants chloroplast (plastome)-encoded antibiotic resistance plays an important role in plant breeding experiments. This character can be used as a marker in designing selection schemes for somatic hybrid recovery (Medgyesy *et al.*, 1980) and organelle behaviour during transfer and interaction in cybrids in *Nicotiana* (Manczel *et al.*, 1986; Medgyesy, 1990).

Available chloroplast markers in higher plants include resistance to the antibiotics streptomycin and lincomycin (Maliga, 1984), triazine herbicides (Arntzen and Duesing, 1983; Cseplo *et al.*, 1985), tentoxin (Durbain and Uchytil, 1977) and maternally inherited pigment deficiency (Kirk and Tilney-Basset, 1978). Of these markers,antibiotic-resistance mutations have been shown to be selectable in cell culture (Maliga, 1984) and therefore suitable for the recovery of chloroplast recombinants.

Recent advances in molecular analysis of the chloroplast genome and the development of protoplast fusion methodologies for plants (Galun, 1982) have proved the way to a more thorough understanding of plastome genetics and nuclear genome/plastome interactions.

The value of plastome-encoded antibiotic resistance markers for studies on organelle inheritance and interactions between nuclear and cytoplasmic genomes in higher plants have been alluded to by a number of workers (Fluhr *et al.*, 1985; Cseplo *et al.*, 1985; Cseplo and Maliga, 1989).

Successful isolation of specific mutants probably requires a plastome-targetted mutagenic agent and a selection scheme that allows sorting out of the multicopy plastome in the somatic dividing cells. Plastome mutations result from either large scale deletions of pt DNA (Day and Ellis, 1984; 1985) or from point mutations (Hosticka and Hanson, 1984; Svab and Maliga, 1986). Mutagenesis followed by selection for specific plastome encoded traits *in vitro* were studied in Solanaceous species including egg plant, red pepper and tobacco (Fluhr *et al.*, 1985; Svab and Maliga, 1986; To *et al.*, 1989; Svab *et al.*, 1990; Sadanandam and Farooqui, 1991; Subhash *et al.*, 1996; Rao *et al.*, 1993, 1997; Ashfaq Farooqui *et al.*, 1997).

Induction of streptomycin-resistant plantlets showing chloroplast encoded mutants have to date been obtained mostly in Solanaceous plants through mutagenesis using alkylating agents like NMU (N-nitroso-N-methyl urea), NEU and EMS (ethyl methane sulfonate) (Fluhr *et al.*,1985; Svab and Maliga, 1986; Svab *et al.*, 1990; Jansen *et al.*, 1990; To *et al.*, 1992; Chen *et al.*, 1993; Rao *et al.*, 1993; Subhash *et al.*, 1996).

The efficiency of mutagenic agents (gamma rays and EMS) for inducing antibiotic resitance has been reported by Rama Swamy *et al* (2005a) in *S. surattense*.

Selection of Streptomycin Resistant Mutants

Induction of streptomycin-resistant shoots from mutagenised and non-mutagenised (control) cotyledon explants of *S. surattense* cultured on selective medium containing 750 mg/l streptomycin (Tables 10.1–10.3). Initial tests were conducted using different concentrations of streptomycin (100–750 mg/L) to determine the complete bleaching and suppresstion of shoot bud induction. It was observed that at 750 mg/l streptomycin caused the total bleaching of cotyledon explants and also suppression of shoot bud proliferation. When mutagenised (with 0.1 per cent EMS) cotyledons were cultured on selective medium containing regeneration medium (RM) supplemented with 750 mg/l streptomycin showed the production of green (resistant) shoots (Plate 9a). Whereas

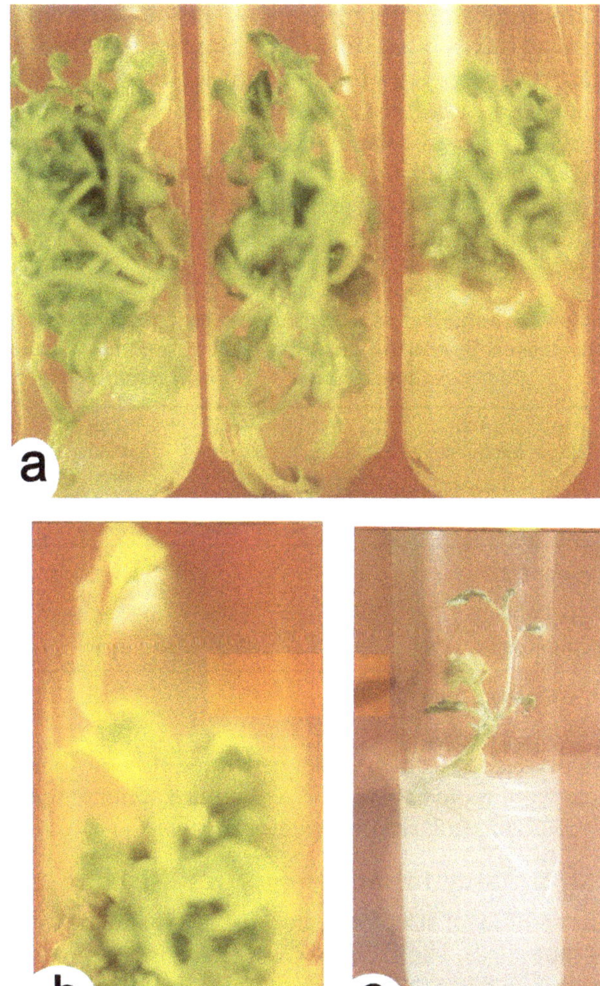

Plate 9a: Streptomycin Resistant Shoots Regenerated in Mutagenised (0.1 per cent EMS) Cotyledon Cultures on MS + 750 mg/L Streptomycin in *S. surattense* (Note the formation of albino shoots).

Plate 9b: Induction of Streptomycin Resistant Shoots in Mutagenised (1 KR) Cotyledon Cultures on MS + 0.5 mg/L IAA + 3.0 mg/L BAP + 750 mg/L streptomycin

Plate 9c: Rooting of Excised Streptomycin Resistant Shoot on MS + 0.1 mg/L IAA + 750 mg/L Streptomycin

streptomycin-resistant shoots were not developed from the non-mutagenised cotyledon explants (Rama Swamy *et al*, 2005a). Gamma-ray irradiated cotyledons induced the high frequency of resistant shoots with more number of resistant shoots (5.6 ± 0.28) per explant on selective medium. Chlorophyll deficient (albino) shoots were also produced along with the resistant shoots in both the types of mutagenised (EMS/g-rays) explants cultured on selective medium but high frequency of albinos were induced in gamma-ray treated explants (Plate 9b).

Table 10.1: Induction of Streptomycin Resistant Shoots from Mutagenised Cotyledon Explants of *S. surattense* Cultured on Regeneration Medium + 750 mg/l Antibiotic

Sl.No.	Treatment and No. of Explants	% of Explants Bleached	% of Explants Forming Albino Shoots	% of Explants with Resistant (Green)	No. of Shoots/ Explant Shoots (S.E)*
1.	Control (70)	100	-	-	-
2.	0.1% EMS (120)	89.6	4.2	6.2	3.8±0.30
3.	1KR-irradiation (120)	62.3	21.2	16.5	5.6±0.28

*: Mean; ±: Standard Error.

Shoots excised from *in vitro* regenerated streptomycin resistant plantlets were rooted on MS basal medium containing 0.1 mg/L IAA and 0.75 mg/mL streptomycin (Plate 9c).

Leaf Strip Assay for Streptomycin Resistance

The stability of streptomycin resistance was studied by a leaf assay (Table 10.2). The explants from the resistant plants cultured on selective medium retained streptomycin resistance, as indicated by the production of green shoots. After three passages of culturing the percentage of explants producing variegated shoots have reduced and the number of green shoots per explant was increased gradually (Rama Swamy *et al.*, 2005a). Thus, the stability of the resistance was increased when the resistant explants were cultured on selective medium containing 750 mg/L streptomycin. The streptomycin-resistant plants were normal with respect to morphology. The ploidy level of streptomycin–resistant was determined and they were found to be diploid (2n =24).

Table 10.2: Leaf Assay for Streptomycin Resistance in *S. surattense*

Passage	% of Explants Producing Variegated Shoots	% of Explants Producing Green Shoots	Average No. of Shoots/Explant (S.E.)*
1	28	72	6.8±0.24
2	10	90	9.2±0.30
3	2	98	12.8±0.2
Control	–	100	18.5±0.43

*: Mean; ±: Standard Error.

Inheritance of Streptomycin Resistance

The transmission of streptomycin resistance to the progeny was investigated as evidence for the mutational origin of streptomycin resistance. The results on inheritance pattern of streptomycin in *S. surattense* is given in Table 10.3. Reciprocal crosses were made between streptomycin-resistant plants (SR) and the original streptomycin-sensitive plants (SS). Plants were scored as streptomycin-resistant when seedlings showed normal growth and rooting on media containing 750 mg/L streptomycin. When the streptomycin-resistant (SR) female parent plant was crossed with the streptomycin sensitive (SS) male parent plant, the offsprings produced were streptomycin-resistant. Whereas the seedlings germinated from the seeds produced from the cross between the streptomycin sensitive female parent and streptomycin resistant male parent, showed that all were stroptomycin sensitive. These results confirm the genetic nature of the streptomycin resistance trait and demonstrated that streptomycin resistance was under the control of maternally (cytoplasmically) inherited mutation *i.e.*, encoded by chloroplast DNA (cp-DNA) (Rama Swamy *et al.*, 2005a).

Table 10.3: The Inheritance of Streptomycin Resistance in *S. surattense*

Cross	% of Germination	Number of Seedings Rested	
		Resistant	Sensitive
SR (♀) × SS (♂)	76	154	-
SS (♀) × SR (♂)	94	-	126

Irradiated cotyledons showed the efficiency over EMS treated explants by developing high frequency of resistant (green) shoots. It was also confirmed that streptomycin resistance was under the control of cp-DNA in *S. surattense* (Rama Swamy *et al*, 2005a).

Fluhr *et al.* (1985) have developed chloroplast encoded antibiotic (streptomycin) resistant mutants in *N. tabacum*. Streptomycin resistance conferred by the streptomycin phosphotransferase (SPT) gene from tn5 is distinct that it provides a screen that is based on color differentiation (Jones *et al.*, 1987; Maliga *et al.*, 1988). The SPT gene has been shown to provide a phenotypic marker in both cell culture and in seedling assays and has been used to detect excision of the AC transposable element in tobacco by a visual assay (Jones *et al.*, 1989). Svab *et al.* (1990) have studied the resistance of spectinomycin and streptomycin in *Nicotiana tabacum* after intorducing aminoglycoside-3-adenyl transferase (aadA genes). The advantage of the chimeric aadA gene, as compared to the SPT gene, is that the same gene confers resistance to both the drugs. In addition, the chimeric aadA gene may be used as a selectable marker (Svab *et al.*, 1990). In *Nicotiana tabacum* also, it was reported that the streptomycin resistance was under the control of chloroplast-DNA as based on co-segregation and marker transfer studies (Maliga, 1984).

According to Etzold *et al.* (1987) and Fromm *et al.* (1989) the chloroplast encoded streptomycin resistance in *Nicotiana tabacum* was caused by point mutations in the chloroplast 16S rRNA or "rps 12" genes (Galili *et al.*, 1989). Similarly, streptomycin resistant is maternally inherited and known to be of chloroplast origin in *Nicotiana plumbaginifolia* (To *et al.*, 1989, 1992). Hsu *et al.* (1993) and Chen *et al.* (1993) have also reported the point mutation at rps 12 in the chloroplast of *N. plumbaginifolia* conferrring streptomycin resistance. Streptomycin binds to the 30s ribosomal subunit of prokaryotic-like ribosomes, inhibits polypeptide synthesis and causes misreading of the genetic code (Davis *et al.*, 1974; Edwards, 1980).

These, streptomycin-resistant mutants have proved to be very useful in designing selection scheme aimed at somatic hybrid recovery in genus *Nicotiana* (Medgyesy *et al.*, 1980).

NMU is the most potent mutagen for inducing mutations in Cp-DNA, resulting a high frequency of cytoplasmic changes

(Hagemann, 1982). EMS and gamma-irradiation induced the highest frequency of plastome mutagenesis in *S. surattense* (Rama Swamy *et al*, 2005a). Similarly, it was also reported in *Solanum melongena* (Sadanandam and Farooqui, 1991; Rao *et al.*, 1993). The successful isolation of mutants was dependent upon establishing conditions, where bleaching was not accompanied by severe growth limitation, which relates to kiloploid nature of the chloroplast genome (Medgyesy, 1990).

Efficient plastome mutagenesis was achieved with N-methyl-N-nitrosoguanide in *N. plumbaginifolia* (To *et al.*, 1989) with EMS in *Capsicum annum* (Subhash *et al.*, 1996) *Solanum melongena* (Rao *et al.*, 1993) and *S. surattense* (Rama Swamy *et al.*, 2005a) with gamma-irradiation in *S. melongena* (Sadanandam and Farooqui, 1991), *S. surattense* (Rama Swamy *et al.*, 2005a) and with NMU in *S. sisymbrifolium* (Rao *et al.*, 1997).

Streptomycin resistance has also been shown to be the result of a recessive mutation in the nucleus (Maliga, 1981). Whereas in *S. surattense* the induced streptomycin resistance was non-mendelian and maternal. Similar observations were also made in a number of species using mutagens like NMU, EMS and gamma-rays. Mc Cabe *et al.* (1989) reported the maternal expression of cp-DNA encoded streptomycin in *Lycopersicon peruvianum, Solanum nigrum* and *Nicotiana tabacum* using nitroso-methyl urea (NMU).

In conclusion streptomycin-resistant Cp-DNA encoded mutants *in vitro* were developed using EMS and gamma-rays in *S. surattense* (Rama Swamy *et al.*, 2005a). These mutants might prove to be of great value in the development of selection systems for somatic hybrids in *Solanum* species.

Chapter 11

Genetic Transformation

Genetic engineering can be vital for an agrarian country like India. It can help in minimizing crop damage through disease and pest resistant varieties, reducing the use of chemicals, enhancing stress tolerance in crop plants thus permitting productive farming on unproductive lands etc. Progress in plant genetic engineering has been spectacular since the recovery of the first transformed plants in the early 1980s. Molecular techniques have now been applied to an array of species, resulting in the generation of numerous transgenic plants. These plants were initially transformed with marker genes, but subsequently with commercially important genes including those enabling agronomic improvement, easier processing and alternative uses (Christou, 1996).

For genetic transformation, DNA can be introduced into a cell using two methods: (i) Direct DNA delivery (ii) Indirect DNA transfer using biological carriers such as *Agrobacterium* or virus (Potrykus, 1990). *Agrobacterium tumefaciens* mediated genetic transformation is the simplest and most reliable available means for introducing foreign DNA into the genomes of dicotyledonous plants (Corbin and Klee, 1991). Plant transformation, via *Agrobacterium tumefaciens,* is usually performed with binary vectors. Most of the available binary vectors contain within the T-DNA (which is transferred to the plant genome) components not required for the intended modification.

The *Agrobacterium* binary vector, pBin 19 has been widely used for introducing foreign genes into plant species (Bevan, 1984). Features such as a positive selection (Km^R) for recovering transformants, multiple cloning site for introducing genes of interest, Lac Z complementation for screening recombinants, and the small size of this plasmid make it ideal for plant transformation experiments.

Agrobacterium mediated genetic transformation plays a great role in transferring agronomically important genes *viz.*, insect, pest, fungal, abiotic stress resistance in to a number of crop plants.

A number of methods have been developed for plant genetic transformation, including particle acceleration (Sanford, 1990), electroporation and polyethylene glycol permeabilization of protoplasts (Potrykus, 1990) and *Agrobacterium* mediated DNA/gene transfer. Although transformation frequencies are remarkably high, the DNA integration pattern is complex in direct DNA delivery system. The present advantages of *A. tumefaciens* mediated transformation are not only the high transformation frequencies obtained but, importantly, the simple integration pattern of the transferred DNA that guarantees practicability (Meyer *et al.*, 1988). With the discovery of the natural ability of *A. tumefaciens* to transfer and integrate part of its DNA into the genome of plants (Gelvin, 1990), various strategies were developed for the exploitation of *A. tumefaciens* strain harbouring Ti-plasmid containing the genes of interest in the T-DNA, allowing transfer to occur, then regenerating transformed plants.

All the *Agrobacterium* mediated transformation methods are based on the observation that the DNA segment/gene transferred by the bacterium can be replaced with foreign DNA/gene, provided transfer signals are present (Hooykaas and Schilperoort, 1992).

Agrobacterium tumefaciens mediated genetic transformation was reported in number of crop plants and also in Solanaceous species. The overall efficiency of plant transformation is generally estimated by the number of transformed plants obtained from infection of a given number of explants. This efficiency is dependent on a number of steps (Higgins *et al.*, 1992). Firstly genes are transferred from *Agrobacterium* to plants during a period of co-culture with plant tissue. Secondly the transformed cells have to be selected (generally by exposure to antibiotics such as kanamycin) and thirdly a tissue culture step is required to regenerate the transformed cells into plants. All these processes must be optimized if transformation is to be efficient. One of the prerequisites for successful gene transfer in to plants is the availability of a suitable protocol for transformation which is compatible with *in vitro* plant regeneration methods of the target plant species.

Although, *Solamun malongena* and its related species have been transformed using *Agrobacterium* mediated transformation by introducing different agronomically important traits (see review Collonnier *et al.*, 2001, Kashyap *et al.*, 2003), but there is no work on *S. surattense.* For the first time the genetic transformation experiments have been carried out in the species.

The transformation has been achieved by using *A. tumifaciens* strain LBA 4404 containing a binary vector (pBIN 19) with β-glucuronidase (GUS) and neomycin phosphotransferase (NPT-II) marker gene cassette in *S. sarattense* (Ramaswamy *et al.*, Pers. Comm.).

The explants (cotyledon and leaf discs) co-cultivated with *A. tumefaciens* formed shoots after 4 weeks of culture on selective shoot regeneration induction medium (MSI). The control explants which were not co-cultivated didn't produce when cultured on MSI medium indicating the effective level of kanamycin (100mg/L).

Kanamycin sensitivity of cotyledon and leaf explants was assessed prior to *Agrobacterium* transformation, to determine the concentration of kanamycin needed for effective growth of transgenic plants. At 50 mg/L kanamycin caused chlorosis and eventual necrosis in all the explants by the end of fourth week. Whereas concentrations of 75 mg/L and 100 mg/L kanamycin completely inhibited the formation of shoot buds and higher concentration of kanamycin (100 mg/L) was used for selection of transformants to prevent possible escapes in *S. surattense* (Ramaswamy *et al.* Pers. Comm.).

High percentage of cultures producing green shoots was observed in cotyledon explants compared to leaf explants (Table 11.1). After 4 weeks, the growing green shoots (Plate 10a,b) from MSI medium were excised and transferred onto MS2 medium for proliferation.

The transformed shoots were multiplied by culturing on MS2 medium containing 100 mg/L kanamycin to confirm the stability of

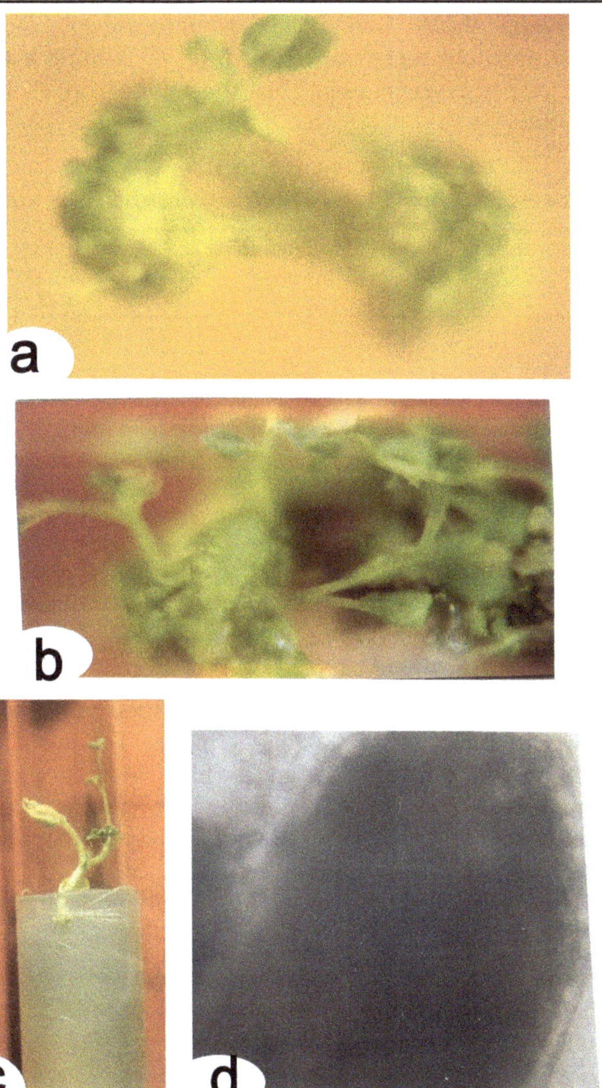

Plate 10a: Kanamycin Resistant Shoots Developing from Cotyledon Explants in *S. surattense*

Plate 10b: Kanamycin Resistant Shoots Developed from Leaf Explants

Plate 10c: *In vitro* Rooting from Transgenic Shoot

Plate 10d: GUS Expression in Transformed Tissue

the transgenic shoots. Leaf explants from transgenic plants when cultured on shoot regeneration medium containing kanamycin showed the plant regeneration. Thus, the stability was also achieved by leaf strip assay.

TABLE 11.1
Agrobacterium* Mediated Genetic Transformation in *S. surattense

Explants	No. of Explants co-cultivated	% of Explants Responding	% of Cultures with Green Shoots	Total No. of Shoots Showing GUS Expression
Leaf	120	45	35	38
Cotyledon	130	55	40	45

Cotyledon: 33 per cent of transformation efficiency.

Leaf: 27 per cent of transformation efficiency.

Most of the transgenic clones appeared morophologically normal in comparison with the untransformed shoots which attained 2-3 cm in length were excised and then transferred to the MS3 medium for rooting (Plate 10c).

The leaves from transgenic shoots were subjected to *in situ* GUS assay. The expression of *uid* A gene was verified by histochemical staining of the leaf of the transgenic plants. The npt II positive regenerants showed the typical indigo blue colouration of X-Gluc treatment (Plate 10c) while the untransformed ones didn't show GUS activity. More than 33 per cent and 27 per cent of the regenerants from cotyledon and leaf explants respectively were GUS positives. These results clearly demonstrate the stability of the transformed plants.

The successful transformation was also reported in a number of other Solanaceous species: *Solanum melongena* (Filippone and Lurquin, 1989; Rotino and Gleddie, 1990; Leone *et al.*, 1993; Fari *et al.*, 1995b), *S. sisimbrifolium* (Rao *et al.*, 1997), *S. muricatum* (Atkinson and Gardner, 1991), *S. tuberosum* (Sheerman and Bevan, 1988; Hoekema *et al.*, 1989); *Lycopersicon esculentum* (Hood *et al.*, 1986; Fillatti *et al.*, 1987; Davis *et al.*, 1991; Frary and Earle, 1996; Rama swamy *et al.*, 2000); *Capsicum annuum* (Liu *et al.*, 1990); *Nicotiana tabacum* (Hood *et al.*, 1986; Rama Swamy *et al.*, 2001) and produced the transgenic plants.

Transformation efficiency was found to be higher in cotyledon explants (33 per cent) compared to leaf explants in *S. surattense.* This transformation efficiency is dependent on various factors : type of explant, explant size, explant orientation on selective regeneration medium, gelling agent and plate sealant and the frequency of transfer on to fresh selective medium. Frary and Earle (1996) have examined the effect of various factors on efficiency of *Agrobacterium*-mediated transformation in *Lycopersicon esculentum* cv. Money maker using cotyledon and hypocotyl explants. Mc Cormick (1991) has reported that the cotyledons were more efficient in generating transgenic shoots as observed in *S. surattense.* Whereas Liu *et al.,* (1990) have reported that the transformation efficiency was higher in leaf followed by hypocotyl and cotyledon explants and also noted the same differential response between the *A. tumefaciens* strains C58 and A281 used in bell pepper. The strain C58 showed more transformation efficiency compared to A281 in all explants co-cultivated. Although most published protocols report the use of whole cotyledons as explants (Davis *et al.,* 1991; McCormick, 1991; Van Rockel *et al.,* 1993), cutting cotyledons into two or three pieces depending on their size is recommended as a way to maximize the number of transformants obtained from a minimum number of seedlings (Frary and Earle, 1996). Armstead and Webb (1987) found that cotyledons of *Lotus corniculatus* were more readily transformed than leaves from seedlings grown *in vitro.* Young leaves were transformed more frequently than old leaves. However, leaves from old papaya plants were found to be more easily transformed than cotyledons or leaves from young plants (Pang and Sanford, 1988).

Agrobacterium-mediated genetic transformation efficiency is not only genotype dependent but also varies with the strains used. The stable integration of GUS and NPT-II genes in *Mentha arvensis* and *M. spicata* has been achieved by *A. tumefaciens* mediated gene transfer. Differences in transformation efficiency between *M. spicata* and *M. arvensis* became apparent in the percentage of explants producing kanamycin resistant (Km^R) calli for the two *Agrobacterium* strains used. *M. spicata* explants produced 53 per cent Km^R calli with GV 2260/GI and 71 per cent with EHA 105/MOG whereas 5 per cent and 1.2 per cent *M. arvensis* Km^R calli were obtained respectively with GV 2260/GI and EHA 105/MOG (Diemer *et al.,* 1999). Likewise, the strain dependent transformation efficiency was also noted by Liu *et al.,* (1990) in bell pepper. Similarly, Davis *et al.,* (1991) reported the difference in degree of transformation caused by three different *A. tumefaciens* strains, pTi-A_6 related plasmids (*i.e.* those in A_6 and A_{66}) have less expression of *vir* genes than pTiBO542 plasmids (*i.e.* those in strain A281). Hussain *et al.*, (1997) also reported in two varieties of chick pea (6153 and CM72), *Agrobacterium* strain A281 was found to be more efficient in transformation than C58.

Venkatachalam *et al.*, (2000) have also found an important factor for efficient transformation in *Arachis hypogaea* was the two day pre-culture of the cotyledon explants, which probably served to reduce wound stress and increased the number of competent cells at the wound site.

Davis *et al.*, (1991) have studied the effect of tomato cultivar, leaf and bacterial strain and density of bacterial inoculum on transformation by *Agrobacterium tumefaciens* and reported the variation in transformation frequencies based on those factors. Plant cultivar was also found to be an important factor in determining transformation frequencies in moth bean in which one cultivar was found to have an 85-fold higher transformation rather than another.

Petunia nurse cell culture technique also enhances the efficiency of *A. tumefaciens* mediated transformation. Recently Rama Swamy *et al.*, (2006d) have reported the high transformation efficiency (96 per cent) with Petunia nurse cell suspension feeder layer culture compared to without feeder layers in leaf discs of *Nicotiana tabacum* cv. Sunsun. Acetosyringone (AS) is a phenolic compound released by wounded cells and it plays an important role in the natural infection of plants by *A. tumefaciens* as it activates the virulence genes of the Ti-plasmid and initiates the transfer of the T-DNA region to the plant DNA. Exogenous addition of AS in the medium has shown to increase Ti transformation frequencies with *Allium cepa, Antirrhinum majus, Arabidopsis thaliana, Atropa belladonna, Brassica compestris, Glycine max, Nicotiana tabacum* and pickling cucumber (Godwin *et al.*, 1991; Sarmento *et al.*, 1992; Frary and Earle 1996) have also reported the enhanced transformation in tomato when the explants (cotyledons and hypocotyl segments) co-cultivated in the presence of AS showed and NPTII gene more efficient expression than control transformants. Further more, regeneration efficiency from transformed explants of *Solanum melongena* was enhanced by using growth regulators, such as TDZ and antibiotics like augmentin (300 µg/ml) and Kanamyan (50 µg/ml) (Billings *et al.*, 1997).

Agronomically important characters have been genetically engineered in major crop plants using *Agrobacterium* mediated genetic transformation. Hinchee (1998) first time achieved the successful recovery of transformed soybean plants for engineering herbicide resistance. Afterwards, this transformation technology was used for introducing agronomically important traits for improvement of the crop in the following species using *Agrobacterium:* sugar beet (Herbicide tolerance–D'Halluin *et al.*, 1995), cotton (Insect resistance, Herbicide tolerance–Umbeck 1987), Papaya (virus resistance–Fitsch *et al.*, 1993), Peanut (Virus resistance–Brar *et al.*, 1994), Poplar (Herbicide resistance–Fillatti, 1988), potato (Insect resistance, virus resistance, herbicide tolerance,–Van den Elzen *et al.*, 1993; Huisman *et al.*, 1992). Squash (virus resistance–Tricoli *et al.*, 1995), egg plant (Bt-cry IIIB–Chen *et al.*, 1995; Arpaia *et al.*, 1997, Billings *et al.*, 1997; Bt-cryIIIA-Hamilton *et al.*, 1997; Bt-cry1Ab-Kumar *et al.*,1998; Parthenocarpic–Rotino *et al.*, 1997) and tomato (Delayed ripening/ increased shelf life), virus resistance–(Sanders *et al.*, 1992; Redenbaugh *et al.*, 1993). Kemper *et al.*, (1992) have developed the transgenic *Arabidopsis thalliana* which are methatrexate resistant due to integration of T-DNA vectors containing a chimeric dihydrofolate reductase gene. Acciarri *et al.*, (2000) have developed the transgenic eggplant conferring resistance against CPB (Clorado potato beetle) natural infectations using Bt-cry IIIB genes. They observed in field trials that the control of CPB infestations. Frijters *et al.*, (2000) have engineered the root-knot nematode *M.incognita* resistance in eggplant by inserting the Mi-1 gene from tomato. More recently eggplant transgenics tolerant to abiotic stresses like salinity, drought and chilling have been achieved by introduction of bacterial mannitol-1-phospho dehydrogenase (mtlD) (Prabhavati *et al.*, 2002). Lawrence and Koundal (2001) have developed the transgenic pigeonpea resistant to chewing insects mainly pod borers using *Agrobacterium tumefaciens* strain GV 2260 containing the construct of isolated cowpea protease inhibitor gene (pCPI).

The Indian Council of Agricultural Research (ICAR) has developed biotech plant types of cotton, brinjal and tomato and now working on evolving similar plants of rice, chickpea and pigeon pea (BIOTECH, 2001). Thus, over 35 genetically improved plants created this way. Trials have also been going on in various laboratories to introduce important traits such as herbicide tolerance, viral resistance, antibiotic stress and disease resistance including 'nif' genes transfer into cereals *viz.*, maize, sorghum and oryza using this *A.tumefaciens.*

After transfer of the gene of interest using *A. tumefaciens* the transgene expression is also an important one. The transgene expression in transgenic populations can vary due to dominant effects exerted by neighbouring plant sequences such as enhancers and silencers may also influence the activity of the introduced genes. Breyene *et al.* (1992a) have studied the influence of the T-DNA configuration on inter-transformant expression variability of a reporter gene. The transcriptional interference can diminish the activity of a gene located downstream in opposite orientation (Ingelbrecht *et al.*, 1991). Breyene *et al.* (1992a) have introduced an additional 3′ *nos* region between the transgene and the RB (right border) in such an orientation that it would stop possible transcripts coming from the flanking plant DNA. The presence of 3′ nos resulted in transgenic population with 1.5 to 2-fold higher mean gus 'A' expression. They have also concluded that the most likely the close proximinity of the '35S' enhancer sequences to 'Pnos' results in an increase of gus 'A' expression. Other molecular causes such as the local DNA structure and/or the higher–order chromatin arrangement (Breyene *et al.*, 1992b) possibly also have an important role in the overall level of gene expression.

In view of the importance of *A. tumefaciens* mediated genetic transformation, the protocol which was developed during the present studies can be utilized to transfer genes of interest for genetic improvement of medicinally important herb *S. surattense.* Genetic engineering studies in *S. surattense* need to be further exploited for the introduction of new genes.

Bibliography

Abraham, A., Krishnan, P.N. and Seenoi, S. (1995). Induction of androgenesis, callus formation and root differentiation in anther culture of cassava (*Manihot esculenta*, Crantz.). *Indian J. Exptl. Biol.,* **33:** 186-199.

Acciarri, N., Vitelli, G., Arpaia, S., Mennella, G., Sunseri, F. and Rotino, G.L. (2000). Field protection against Colarado Potato Beetle of Bt-expressing in transgenic explant. *Hort. Sci.*

Ahuja, A., Verma, M. and Grewal, S. (1982). Clonal propagation of *Ocimum* species by tissue culture. *Indian J. Exptl. Biol.,* **20:** 455-458.

Alia, Hayashi, H., Sakomoto, A. and Murata, N. (1998). Enhancement of the tolerance of *Arabidopsis* to high temperature by genetic engineering of the synthesis of glycine betaine. *Plant J.,* **16:** 155-160.

Amberger, L.A., Palmer, R.G. and Shoemaker, R.C. (1992). Analysis of culture induced variation in soybean. *Crop. Sci.,* **32:** 1103-1108.

Ammirato, P.V. (1977). Hormonal control of somatic embryo development from cultured cell caraway. *Plant Physiol.,* **59:** 579-586.

Ammirato, P.V. (1983a). Embryogenesis. In: *Hand book of plant cell cultures* (eds.) Evans, D.A., Sharp, W.R., Ammirato, P.V. and Yamada, Y. McMillan Publishing Co., New York, pp. 82-123.

Ammirato, P.V. (1983b). The regulation of somatic embryos development in plant cell cultures, suspension cultures technique and hormone requirements. *BioTechnol.,* **1:** 68-74.

Ammirato, P.V. (1987). Organization events during somatic embryogenesis. In: *Plant Tissue and Cell Culture,* (Eds.) Green, C.E., Somers, D.A., Hackett W.P. and Biesboer, D.D. Alan R. Liss, Inc. New York, pp. 57-81.

Anand, P.H.M. and Hirahardan, M. (1997). *In vitro* multiplication of greater galangel (*Alpinia galanga* (L.) Wild) medicinal plant. *Phytomorphology,* **47:** 45-50.

Aonymous, 2006. www.google.com

Anwar, S. and Siddiqui, S.A. (2000). *In vitro* organogenesis in *Ocimum sanctum* L.–A multipurpose Herb. *Phytomorphology,* **50:** 27-35.

Armstead, I.P. and Webb, K. (1987). Effect of age and type of tissue on genetic transformation of *Lotus corniculatus* by *Agrobacterium tumefaciens. Plant Cell Tiss. Organ Cult.,* **9:** 95-101.

Arntzen, C.J. and Duesing, J.H. (1983). Chloroplast encoded herbicide resistance. In: *Advances in gene technology: molecular genetics of plants and animals.* (eds.) Downey, K. Voellmy, R.W., Ahmad, F., Schut, Z.J., Academic Press, New York, pp. 273-292.

Arpaia, S., Mennella, G., Onofaro, V., Perri, E., Sunsari, F. and Rotino, G.L. (1997). Production of transgenic egg plant (*Solanum melongena* L.) resistant to colorado potato beetle L. (*Leptinotarsa decemlineata* Say). *Theor. Appl. Genet.,* **95:** 329-334.

Asao, H., Tanigawa, M., Okayama, K. and Arai, S. (1992). Breeding of resistant *Solanum* spp. for bacterial wilt by cell selection using a wilt inducing product. *Bull. Nara. Agr. Exp. St.,* **23:** 7-12.

Ashalatha, S.N. and Seo, B.B. (1993). Plantlet regeneration from callus initiated from flower buds in the wild species *Allium senescens* var. *minor*. *Plant Cell Tissue Org. Cult.,* 47: 205-207.

Ashfaq Farooqui, M., Rao, A.V., Jayasree and Sadanandam, A. (1997). Induction of atrazine resistance and somatic embryogenesis in *Solanum melongena*. *Theor. Appl. Genet.*, **95:** 702-705.

Atkinson, R.G. and Gardner, R.C. (1991). *Agrobacterium* mediated transformation of pepino and regeneration of transgenic plants. *Pl. Cell Rep.*, **10:** 208-12.

Atree, S.M. and Fowke L.C. (1993). Embryogeny of gymnosperms-advances in synthetic seed technology. Conifers. *Plant Cell Tissue Org. Cult.*, **35:** 1-35.

Azad, M.A.K. and Amin, M.N. (1998). *In vitro* regeneration of plantlets from internode explants of *Adhatoda vasica* Nees. *Pl. Tiss. Cult.*, **8:** 27-34.

Baburaj, S. and Gunasekaran, K. (1994). Regeneration of plants from leaf callus cultures of *Solanum pseudocapsicum* L. *Indian J. Expt. Biol.*, 141-143.

Baily, L.H. (1949). *Mannual of cultivated plants*. The Macmillan Co., New York.

Bais, H.P., George, J. and Ravi Shankar, G.A. (2000). *In vitro* propagation of *Decalepis hamiltonia* Wight and Arn. an endangered shrub through axillary bud cultures. *Curr. Sci.*, **79:** 408-410.

Bajaj, Y.P.C. and Gupta, R.K. 1987. Plants from salt tolerant cell lines of Napier grass, *Penisetum purpureum* Schcum. *Indian J. Exptl. Biol.*, **25:** 58-60.

Bajaj, Y.P.S. (1983). *In vitro* production of haploids. In: *Handbook of Plant Cell Culture.* (eds.) Evans, D.A., Sharp, W.R., Ammirato, P.V. and Yamada, Y., **Vol. I**, Macmillan New York, pp. 228-287.

Bajaj, Y.P.S. (1988). Some Indian ornametal trees: *Cassia fistula* Linn., *Poinciana regia* (Boj.) and *Jacaranda acutifolia* Auct. In: *Biotechnology in Agriculture and Forestry*, 5, Springer Berlin, Heidelberg, New York.

Baker, C.M. and Wetzstein, H.Y. (1994). Influence of auxin type of concentration on peanut somatic embryogenesis. *Plant Cell Tissue Org. Culture*, **35:** 151-156.

Balachandran, S.M., Bhat, S.R. and Chandel, K.P.S. (1990). *In vitro* clonal multiplication of turmeric (*Curcuma* spp.) and ginger (*Zingiber officinale* Rose). *Plant Cell Rep.*, **8:** 521-524.

Banerjee, S. and Gupta, S. (1975). Embryoid and plantlet formation, from stock cultures of *Nigella* tissues. *Physiol. Plant.,* **34:** 243-245.

Bapat, U.A. and Rao, P.S. (1977). Experimental control of growth and differentiation in organ cultures of *Physalis minima* Linn. *Z. Pflanzen. Physiol.,* **85:** 403-416.

Bartels, D., Hanke, C., Schneider, K., Michel, D. and Salamini, F. (1992). A desiccation-related Elip-like gene from the resurrection plant *Craterostigma plantagineum* is regulated by light and ABA. *EMBO Journal,* **8:** 2771-2778.

Beerasft, P.W. and Taylor, G.A. (1992). Genetic variation for anther culture *callus inducibility* in crosses of highly culturable winter wheats. *Plant Breeding,* **108:** 19-25.

Bendich, A.J. (1982). In: *Mitochondrial genes.* (eds.) Slonim Ski. P. and Atterdi, G., Cold Spring Harbour Lab., pp. 477-481.

Ben-Hayyim, J. and Neumann, H. (1983). Stimulatory effects of glycerol on growth and somatic embryogenesis in *Citrus* callus cultures. *Z. Pflanzen. Physiol.,* **110:** 331-337.

Bevan, M. (1984). Binary *Agrobacterium* vectors for plant transformation. *Nucl. Acids. Res.,* **12:** 8711-8721.

Bhaskar, P. and Subhash, K. (1996). Micropropagation of *Acacia mangium* Wild. through nodal bud culture. *Indian J. Exptl. Biol.,* **34:** 590-591.

Bhaskaran, S., Rigoldi, M. and Smith, R.H. (1992). Developmental potential of *Sorghum* shoot apices in culture. *J. Plant Physiol.,* **140:** 481-484.

Bhat, S.K., Kachar, A. and Chandel, K.P.S. (1992). Plant regeneration from callus cultures of *Piper longum* L. by organogenesis. *Plant Cell Rep.,* **11:** 525-528.

Bhatt, P.N., Bhat, D.P. and Sussex, I.M. (1979). Organ regeneration from leaf discs of *Solanum nigrum, S. dulcamara* and *S. khasianum. Z. Pflanzenphysiol.,* **95:** 355-362.

Bhatt, S.R., Chandel, K.P.S. and Malik, S.K. (1995). Plant regeneration from various explants of cultivated *Piper* species. *Plant Cell Rep.,* **14:** 398-402.

Bhattacharya, S. and Bhattacharya, S. (1997). Rapid multiplication of *Jasminum officinale* L. by *in vitro* culture of nodal explants. *Plant Cell Tissue Org. Cult.,* **51:** 57-60.

Bhattacharjee, S.K. 1998. *Hand Book of Medicinal Plants-1,* Pointer Publishers, Jaipur, pp. 326.

Bhojwani, S.S. (1980). Micropropagation method for a hybrid Willow (*Salix mastudan X alba* NZ 1002). *N. Z. J. Bot.,* **18:** 209-214.

Bhojwani, S.S. and Razdan, M.K. (1996). *Plant Tissue Culture Theory and Practice,* Elsevier, Amsterdam pp. 125-166.

Biggs, B.J., Smith, M.K. and Scott, K.J. (1986). The use of embryo culture for the recovery of plants from Cassaova (*Manihot esculenta* Grantz.) *Seeds. Plant Cell Tissue Org. Cult.,* **6:** 229-234.

Billings, S., Jelen Kovic, G., Chin, C.K. and Eberthadt, J. (1997). The effect of growth regulation and antibiotics on egg plant transformation. *J. Amer. Soc. Hort. Sci.,* **122:** 158-162.

Binch, D.Q., Heszky, L.E., Gyulai, G. and Grillag, A.C. (1992). Plant regeneration of NaCl–pretreated cells from long term suspension culture of rice (*Oryza sativa* L.) in high saline conditions. *Plant Cell Tiss. Org. Cult.,* **29:** 75-82.

Binzel, M.I., Sankhla, N., Sangeeta Joshi and Sankhla, D. (1996). Induction of direct somatic embryogenesis and plant regeneration in pepper (*Capsicum annuum* L.). *Plant Cell Rep.,* **15:** 536-540.

Binzel, M.L., Hasegawa, P.M., Handa, A.K. and Brassan, R.A. (1985). Adaptation of tobacco callus to NaCl. *Plant Physiology,* **79:** 118-125.

Biotechnology (2001). **3:** 1-4, *Mansanto,* Mumbai.

Bjornastad, A., Skinners, H. and Thoresen, K. (1993). Comparison between doubled haploid lines produced by anther culture– the *Hordeum bulbosum*-mediated lines produced by single seed descent in barley crosses. *Euphytica,* **66:** 135-144.

Bohnert, H.J. and Jensen, R.G. (1996). Strategies for engineering water stress tolerance in plants. *Trends Biotech, 14:*89-97.

Bohnert, H.J. and Shen, B. (1999). Transformation and compatible solutes. *Scien, Hort., 78:* 237-260.

Bottino, P.J. (1975). The potential of Genetic manipulation of plant cell cultures for plant breeding. *Rad. Bot.,* **15:** 1-16.

Brar, G., Cohen, B.A., Vick, C.L. and Johnson, G.W. (1994). Recovery of transgenic peanut (*Arachis hypogaea* L.) plants from elite cultivars, utilizing ACCell Technology, *Plant J.,* **5:** 745-753.

Breyen, P., Gheysen, G., Jacobs, A., Van Montagu, M. and Depicker, A. (1992a). Effect of T-DNA configuration on transgene expression. *Mol. Gen. Genet.,* **235:** 389-396.

Breyen, P., Van Montagu, M., Depicker, A. and Gheysen, G. (1992b). Characterization of a plant scaffold attachment region in a DNA fragment that normalizes transgene expression in tobacco. *Plant Cell Rep.,* **4:** 463-471.

Broglie, K., Chet, I., Holliday, M., Crossman, R., Biddle, P., Knowlton, S., Manain, J. and Broglie, R. (1991). Transgenic plants with enhanced resistance to fungal pathogen *Rhizoctonia solani. Science,* **254:** 1194-1197.

Butcher, D.N. (1977). Plant tumor cells. In: *Plant Tissue and Cell Culture,* (Ed.) Street, H.E. Blackwell Scientific Publishers, USA, pp. 429-461.

Cai, Q., Szarezko, I., Polok, K. and Meluszynski, M.M. (1992). The effect of sugars and growth regulators on embryoid formation and plant regeneration from barley anther culture. *Plant Breed.,* **109:** 218-226.

Cano, A.E., Perez-Alfocea, F., Moreno, V., Caro, M. and Bolarin, M.C. (1998). Evaluation of salt tolerance in cultivated and wild tomato species through *in vitro* shoot apex culture. *Plant Cell Tissue Org. Cult.,* **53:** 19-26.

Caplan, A., Herrara Estrella, L., Inze, D., Van Montagum, M. Schell, J. and Zambryski, P. (1985). Introduction of genetic material into plant cells. *Science,* **222:** 815-821.

Cardoza, V. and D'Souza, L. (2000). Direct somatic embryogenesis from immature zygotic embryos in cashew (*Anacardium occidentale* L.). *Phytomorphology.,* **50:** 201-204.

Carlson, P.S. and Polacco. (1975). The use of protoplasts for genetic research. *Proc. Natl. Acad. Sci., (U.S.A.),* **70:** 598-602.

Cavallini, A.and Natali, L. (1989). Cytological analyses of *in vitro* somatic embryogenesis in *Brimeura amathystina* Salisb (Liliaceae). *Plant. Sci.,* **62:** 255-261.

Chaleff, R.S. (1981). *Genetics of Higher plants. Applications of Cell Cultures,* Cambridge University Press, Cambridge.

Chaleff, R.S. (1983). Isolation of agronomically useful mutants from plant cell cultures. *Science,* **219:** 676-682.

Chand, S. and Singh, A.K. (2001). Direct somatic embryogenesis from zygotic embryos of a timber-yielding Leguminous tree. *Hardwickia binata* Roxb. *Curr. Sci.,* **80:** 882-888.

Chandler, S. and Dodds, J. (1983). Solasodine production in rapidly proliferating tissue cultures of *Solanum lacianatum. Plant Cell Rep.,* **2:** 69-72.

Chandler, S.F. and Thorpe, T.A. (1987). Characterization of growth, water relations, and proline accumulation in sodium sulphate tolerant callus of *Brassica napus* L. cv. Westar (Canola). *Plant Physiol.,* **84:** 106-111.

Chaturvedi, S.B. and M. Sinha (1979). Mass clonal propagation of *Solanum khasianum* through tissue culture. *Indian J. Exptl. Biol.,* **17:** 153-157.

Chaturvedi, S.B., Dey, N.C., Shukla, A.K., Bora, R.K. and Saikra, B.K. (1979). Determination of solasodine from *Solanum khasianum* Clarke. *Indian Drugs,* 153-157.

Chawla, H.S. and Wenzel, G. (1987a). *In vitro* selection of barley and wheat for resistance against *Helminthosporium sativum. Theor. Appl. Genet.,* **74:** 841-845.

Chawla, H.S. and Wenzel, G. (1987b). *In vitro* selection for fusaric acid resistant barley plants. *Plant Breed,* **99:** 159-163.

Chee, P.P. (1992). Initiation and maturation of somatic embryos of Squash (*Cucurbita pepo* L.). *Hort. Sci.,* **27(1):** 59-60.

Chen, M.H., Yang, W.P., Chen, C.C., Lao, Y.K. and Lin, T.Y. (1993). Point mutation in the chloroplast rsp 12 genes from *Nicotiana plunbaginifolia* confers, Streptomycin resistance. *Plant Mol. Biol.,* **23:** 179-183.

Chen, Q., Jelenkovic, G., Chin, C., Billings, S., Eberhardt, J. and Goffreda, J.C. (1995). Transfer and transcriptional expression of coleopteran "*Cry*" IIIB endotoxin gene of *Bacillus thuringiensis* in egg plant. *J. Amer. Soc. Hort. Sci.,* **120:** 921-927.

Chopra, R.N., Chopra, I.C., Handa, K.L., and Kapoor, L.D. (1958). *Indigenouns drugs of India,* 2nd Edition, U.N. Dhur and Sons, Pvt., Ltd., Calcutta, pp. 525.

Christopher, T. and Rajam, M.V. (1996). Effect of genotype, explant and medium on *in vitro* regeneration of red pepper. *Plant Cell Tissue Org. Cult.,* **46:** 245-250.

Christou, P. (1996). *Transformation Technology Trends in Plant Science-1,* 423-431.

Cocking, E.C. (1960). A method for the isolation of plant protoplasts and vacuoles. *Nature,* **187:** 962-963.

Cocking, E.C. (1978). Selection and somatic hybridisation. In: *Frontiers of Plant Tissue Culture.* (ed.) T.A. Thorpe, Academic Press, pp. 151-158.

Collonnier, C., Fock, I., Kashyap, V., Rotino, G.L., Daunay, M.C., Lian, Y., Mariska, I.K., Rajam, M.V., Servaes, A., Ducreux, G. and Sinhachakr, D. 2001. Applications of biotechnology in Egg Plant. *Plant Cell Tissue Org. Cult.,* **65:** 91-107.

Compton, M.E. and Gray, D.J. (1993). *Plant Cell Rep.,* **12:** 61–65.

Corbin, D.R. and Klee, H.J. (1991). *Agrobacterium tumefaciens–* mediated plant transformation system. *Current Options in Biotech.,* **2: 147-152.**

Croughan, T.P., Stavarek, S.J. and Rains, D.W. (1981). *In vitro* development self resistant plants. *Env. Exptl. Bot.,* **21:** 311-324.

Cseplo, A. and Maliga, P. (1989). Large scale isolation of maternally inherited lincomycin resistance mutations, in diploid *Nicotiana plumbaginifolia* protoplast cultures. *Mol. Gen. Genet.,* **196:** 407-412.

Cseplo, A., Medgyesy, P., Hideg, E., Demeter, S., Marton, L. and Maliga, P. (1985). Triazine-resistant *Nicotiana* mutants from photomixotrophic cell cultures. *Mol. Gen. Genet.,* **200:** 508-510.

Cushman, J.C., De Rocher, E.I. and Bohnert, H.J. (1990). Gene expression during adaptation to salt stress. In: *Environmental Injury to Plants* Katterman, F. (ed.), Academic Press, Inc., San Diego, pp. 173-203.

D'Halluin, K. *et al.* (1995). Transformation of sugar beet (*Beeta vulgaris* L.) and evaluation of herbicide resistance in transgenic plants. *Biotechnology,* **10:** 309-314.

Das, T.O. and Mitra, G.C. (1990). Micropropagation of *Eucalyptus tereticornis* Smith. *Plant Cell Tiss. Org. Cult.,* **22:** 95-103.

Davis, B.D., Tai, P.C. and Wallace, B.J. (1974). Complex interactions of antibiotics with the ribosome. In: *Ribosomes.* eds. Nomura, M., Tissueres, A., Lengeys, P. Cold Springer Harbour Laboratory, New York, 771-791.

Davis, M., Linebeerger, R.D. and Miller, A.R. (1991). Effects of tomato cultivar, leaf age and bacterial strain on transformation by *Agrobacterium tumefaciens. Plant Cell Tiss. Org. Cult.,* **24:** 115-121.

Day, A. and Ellis, T.H.N. (1984). Chloroplast DNA deletions associated with wheat plants regenerated from pllen: possible basis for maternal inheritance of chloroplasts. *Cell,* **9:** 671-678.

Day, A. and Ellis, T.H.N. (1985). Deleted forms of plastid DNA in albino plants from cereal anther culture. *Curr. Genet.,* **9:** 671-678.

De Langhe, E. and De Bruijne, A. (1976). Continuous propagation of tomato plants by means of callus culture. *Sci. Hortic.,* **4:** 221-227.

Delauney, A.J. and Verma, D.P.S. (1990). A soybean gene encoding D'-pyrroline-5-carboxylate reductase was isolated by functional complementation in *Escherichia coli* and is found to be osmoregulated, *Mol. and Gen. Genet.,* **221**, 299-305.

Desai, H.V., Bhatt, P.N. and Metha, A.R. (1986). *Plant cell Rep,* **3:** 190-191.

Dhawan, V. (1993). Tissue culture of hardwood species. In: *Plant Biotechnology: Commercial Prospects and Problems.* (eds.) J. Prakash and R.L.M. Pierik, Oxford and IBH, Publishing Co. Pvt. Ltd., New Delhi, pp. 43-71.

Dhawan, V. and Bhojwani, S.S. (1985). *In vitro* propagation of *Leucaena leucocephala* (Lam) de. Milt. *Plant Cell Rep.,* **4:** 315-318.

Diemer, F., Caissard, J.C., Moja, S. and Jullien, F. (1999). *Agrobacterium tumefaciens*–mediated transformation of *Mentha spicata* and *Mentha arvensis. Plant Cell Tiss. Org. Cult.,* **57:** 75-78.

Dix, P.J. and Street, H.E. (1975). Sodium chloride-resistant cultured cell lines from *Nicotiana sylvestris* and *Capsicum annuum. Plant Science Lett.,* **5:** 231-237.

Dorion, N., Wies, N., Burteaux, A. and Bigot. (1999). Protoplast and leaf explant culture of *Lycopersicon cheesmanii* and salt-tolerance of protoplast derived calli. *Plant Cell Tiss. Org. Cult.,* **56:** 16-19.

Dubey, R.S. (1997). *Strategies for improving salt tolerance in Higher plants,* (Eds.), Jaiwal, P.K., Singh, R.P. and Gulati, A. *Science Publishers Inc.,* USA, pp. 129-158.

Duhoux, E. and Davies, D. (1985). Shoot production from cotyledons buds of *Acacia albida* and influence of sucrose on rhizogenesis. *J. Plant Physiol.,* **121:** 175-180.

Dumas de Vaulx, R. and Chambonnet, D. (1982). Culture *in vitro* d' anthers, dhaubergine (*Solanum melongena* L.) stimulation de la production de plantes an moyen, de traitments at +36°C associeas a de faibles teneurs, substances de croissance, *Agronomie,* **2:** 983-988.

Dunstan, D.I. and Short, K.C. (1979). Improved growth of tissue cultures of onion *Allium cepa. Physiol. Plant,* **27:** 70-72.

Dunwell, J.M. (1985). In: *Cereal Tissue and Cell Culture* (eds.) Bright, S.W.J. and Jones, M.J.K., Martinus Nijhoff, The Hague.

Dunwell, J.M. (1986). Pollen, ovule and embryo culture as tools in plant breeding In: *Plant Tissue Culture and its Agriculture Applications,* Mithers LA and Alderson PG (Eds.) (pp. 375-404), Butter Worthys, London.

Durbain, R.D. and Uchytil, T.T. (1977). Cytoplasmic inheritance of chloroplast coupling factor I subunits. *Biochem. Genet.,* **15:** 1143-1146.

Durga, M. Brave and Mehta, A.R. (1993). Clonal propagation of mature elite tree *Cammiphora wightii. Plant Cell Tiss. Org. Cult.,* **35:** 237-240.

Eberhardt, H.J. and Wegmann, K. (1989). Effects of abscisic acid and proline on adaptation of tobacco callus cultures to salinity and osmotic shock. *Physiol. Plant,* **76:** 283-288.

Edwards, D. (1980). *Antimicrobiol drug action,* MacMillion Pvt. Co., London, pp. 193-216.

Ekiz, H. and Konzak, C.F. (1991). Nuclear and cytoplasmic control of anther culture response in wheat: 111. Common wheat Crosses. *Crop. Sci.,* **31:** 1432-1436.

Ekiz, H. and Konzak, C.F. (1993). Effect of different light applications on anther culture response of spring bread wheat (*Triticum aestivum* L.). *Doga Turkish Journal of Agriculture and Forestry,* **17:** 511-520.

Ekiz, H. and Konzak, C.F. (1994). Preliminary dialled analysis of anther culture response in wheat (*Triticum aestivum* L.). *Plant Breeding,* **113:** 47-52.

Emmanuel, S., Ignacimuthu, S. and Kathiravan, K. (2000). Micropropagation of *Wedelia calendulacea* Less. A medicinal plant. *Phytomorphology.* **50:** 195-200.

Etzold, T., Fritz, C.C., Schell, J. and Schreir, P.H. (1987). A point mutation in chloroplast 16s to RNA gene streptomycin resistant *Nicotiana tabacum. FEBS Lett.,* **21:** 343-362.

Evans, D.A. (1989). Somaclonal variation-genetic basis and breeding applications. *Trends in Genetics,* **5:** 46-50.

Evans, D.A. and Bravo, J.E. (1983). Protoplasts, isolation and culture. In: *Handbook of plant cell culture,* Vol. 1 (Eds.) Sharp, W.R., Ammirato, P.V. and Yamada, Y., MacMillan Pub. Co., New York, pp. 124-176.

Evans, D.A., Sharp, W.R., Ammirato, P.V. and Yamada, Y. (1983). *Handbook of plant cell culture,* Vol. 1, Macmillan, New York.

Evans, D.A., Sharp, W.R. and Medina Fillo, H.P. (1984). Somaclonal and gametoclonal variations. *Amer. J. Bot.,* **71:** 759-774.

Evans, D.A. and Sharp, W.R. (1986). Applications of somaclonal variation. *Biotechnology,* **4:** 528-532.

Fari, M. and Czako, M. (1981). Relationship between position and morphogenetic response of pepper hypocotyl explants cultured *in vitro. Sci. Horti.,* **5:** 207-213.

Fari, M., Csanyl, M., Mityko, J., Peredi, A., Szasy, A. and Csillag, A. (1995a). An alternative pathway of *in vitro* organogenesis in higher plants: Plant regeneration via decapitated hypocotyls in three Solanaceous vegetables genera. *Hort. Sci.,* **27:** 9-16.

Fari, M., Nagy, I., Csanyl, M., Mityko, I. and Andrsfalvy, A. (1995b). *Agrobacterium* mediated genetic transformation and plant regeneration via organogenesis and somatic embryogenesis from cotyledon leaves in egg plant (*Solanum melongena* L. cv. *Keiskemefi lila*). *Plant Cell Rep.,* **15:** 82-86.

Fassuliotis, G. (1975). Regeneration of whole plants from isolated stem parenchyma cells of *Solanum sisymbrifolium. J. Amer. Soc. Hort. Sci.,* **100:** 636-638.

Fewell, A.M., Roddick, J.G. and Weissenberg, M. (1994). Interactions between the glycoalkaloids solasonine and solamargine in relation to inhibition of fungal growth. *Phytochem.,* **37:** 1007-1011.

Filho, G.E. and Sodek, L. 1988. Effect of salinity on ribonuclease activity of *Vigna unguiculata* cotyledons during germination. *J. Plant Physiology,* **132:** 307-311.

Filippone, F. and Lurquin, P.F. (1989). Stable transformation of egg plant (*Solanum melongena* L.) by co-cultivation of tissue with *Agrobacterium tumefaciens* carrying a binary plasmid vector. *Plant Cell Rep.,* **8:** 370-373.

Fillatti, J.J. (1988). Development of glyphosate tolerant popular plants through expression of a mutant aro 'A' gene from *Salmonella typhimurium. Genet. Mampwoody Plants,* **44:** 243-249.

Fillatti, J.J., Kiser, J., Rose, R. and Comai, L. (1987). Efficient transfer of a glyphosate tolerance gene into tomato using a binary *Agrobacterium tumefaciens. Vector Biotechnology,* **5:** 726-730.

Finar, J.J. (1994). Plant regeneration via embryogenic suspension cultures. In: *Plant Cell Culture: A Practical Approach,* Dixon, R.A. and Gonzales, R.A. (eds.), Oxford University Press, Oxford, pp. 67-102.

Finner, J.J. (1988). Apical proliferation of embryogenic tissue of soybean (*Glycine max* L. Merill). *Plant Cell Rep.,* **7:** 238-241.

Fischhoff, D.A., Bowdish, K.S., Perlek, F.J., Marrone, P.G., Mc Cormick, S.M., Niedermayer, E.J., Rochester, E.J., Rogers, S.G. and Fray, R.T. (1987). Insect tolerant transgenic tomato plants. *Biotechnology,* **5:** 807-813.

Fitch, M.M.M., Manshardt, R.D., Gonslaves, D. and Slinghthom, J.L. (1993). Transgenic papaya plants from *Agrobacterium*-mediated transformation of somatic embryos. *Plant Cell Rep.,* **12:** 245-249.

Flick, C.E., Evans, D.A. and Sharp, W.A. (1983). Organogenesis. In: *Hand book of Plant Cell Culture*. Vol. 1 (eds.) D.A. Evans, W.R. Sharp, P.V. Ammirato and Y. Yamada, McMillan, New York, pp. 13-81.

Flores, H.E. and Medina-Bolivar, F. (1995). Root cultures and plant natural products unearthing the hidden half of plants metabolism. *Plant Tiss. Cult. & Biotech,* **2:** 59-74.

Fluhr, R., Dvora, A., Galun, E. and Edelman, M. (1985). Efficient induction and selection of chloroplast encoded antibiotic resistant mutants in *Nicotina. Proc. Natl. Acad. Sci.,* **82:** 1485-1489.

Frary, A. and Earle, E.D. (1996). An examination of factors effecting the efficiency of *Agrobacterium* mediated transformation of tomato. *Plant Cell Rep.,* **16:** 235-240.

Frijters, A., Simons, G., Varga, G., Quaedvlieg, N., and deBoth, M. (2000). The Mi-1 gene confers resistance to the root-knot nematode *Meloidogyne incognita* in transgenic eggplant. VIII Conf. Plant and animal genome, San Diego, CA, USA.

Fromm, H., Galun, E. and Edelman, M. (1989). A novel site for streptomycin resistance in the 530 loop of chloroplast 16s ribosomal RNA. *Plant Mol. Biol.,* **12:** 499-505.

Fugita, Y., Hera, Y., Suga, C. and Morimoto, M. (1981). Production of shikonin derivatives by cell suspension cultures of *Lithospermum erythroshizen*. A new medium of the production of shikonin derivatives. *Plant Cell Rep.,* **1:** 61-63.

Galili, S., Fromm, H., Auis, D., Edelman, M. and Galun, E. (1989). Ribosomal protein S_{12} and a site for streptomycin resistance in *Nicotiana* chloroplast. *Mol. Gen. Genet.,* **218:** 289-292.

Galun, E. (1982). In: *Plant Improvements and Somatic Cell Genetics* (eds.) Vasil, I.K., Fr om, K.J. and Scowcroft, W.R., Academic Press, New York, pp. 205-210.

Gamble, J.S. (1986). *Flora of the presidency of Madras*, vol. II, Bishen Singh Mahendrapal Singh, Dehradun.

Gamborg, O.L., Miller, R.A. and Ojima, K. (1968). Nutrient requirements of suspension cultures of soybean root cells. *Exp. Cell Res.,* **50:** 151-158.

Gamborg, O.L., Shyuluk, J.P. and Constable, F. (1977). Morphogenesis and plantlet regeneration from callus cultures of immature embryos of *Sorghum*. *Plant Sci. Lett.*, **10:** 67-74.

Gangopadhyay, G., Basu, S., Mukherjee, B.B. and Gupta, S. (1997a). Effects of salt and osmotic shocks on unadapted and adapted callus lines of tobacco. *Plant Cell Tiss. and Organ Cult.*, **49:** 45-52.

Gangopadhyay, G., Basu, S. and Gupta, S. (1997b). *In vitro* selection and physiological characterization of NaCl–and mannitol-adapted callus lines in *Brassica juncea*. *Plant Cell Tiss. and Organ Cult.*, **50:** 161-169.

Garin, E., Grenier, E. and Granier, G.D. (1997). Somatic embryogenesis in wild cherry (*Prunus avium*). *Plant Cell Tiss. and Organ Cult.*, **48:** 83-91.

Gautheret, R.J. (1939). Sur la possibilitie de realiser la culture indefinie des tissue de tubercules de carotte. *C.R. Acad. Sci.* (Paris), **208:** 118-120.

Gautheret, R.J. (1959). In: *Culture des tissus vegetaux Techniques et. Realisations*. Masson, Paris.

Gautheret, R.J. (1966). In: *Cell differentiation and morphogenesis*, (ed.) Beermann, W. pp. 55-71.

Geetha, S. and Shetty, S.A. (2000). *In vitro* propagation of *Vanilla planifolia*, a tropical orchid. *Curr. Sci.*, **79:** 886-889.

Gelvin, S.B. (1990). Crown gall disease and hairy root disease, a sledge hammer and a tack hammer. *Plant Physiol.*, **92:** 281-285.

Gengenbach, B.G., Green, C.E. and Donovan, C.D. (1977). Inheritance of selected pathotoxin resistance in maize plants regenerated from cell cultures. *Proc. Natl. Acad. Sci.* (USA), **74:** 5113-5117.

George, E.F. and Sherrington, A.D. (1984). Plant propagation by tissue culture. In: *Hand book and directory of commercial laboratories*, Gastern Press, Reading Berks, pp. 55.

George, P.S., Ravishanker, G.A. and Venkataramana, L.V. (1993). Clonal multiplication of *Gardenia jasminoides* Ellis through axillary bud culture. *Plant Cell Rep.*, **13:** 59-62.

Gharyal, P.K. and Maheswari, S.C. (1982). *In vitro* differentiation of plantlets, from tissues culture of *Albizzia lebbeck* (L.). *Plant Cell Tiss. and Organ Cult.*, **2:** 49-53.

Gill, R.I.S., Gill, S.S. and Gosal, S.S. (1996). Protocol for *in vitro* asexual multiplication of *Azadirachta indica.* In: *Plant Tissue Culture* (eds.) Islam, A.S., Oxford and IBH Publishing Co., pp. 155-160.

Giulietti, A.M., Nigra, H.M. and Caso, O. (1991). *Solanum eleagnifolium* (Silver Leaf Nightshade). *In vitro* culture and the production of Solasodine. In: *Biotechnology in Agriculture and Forestry,* Vol. 15, Medicinal and Aromatic Plants III. (ed.) Bajaj, Y.P.S., pp. 432-450, Springer-Verlag, Berlin.

Gleba, Y.Y. and Shlumukov, B.R. (1990). Selection of somatic hybrids. In: *Plant Celll Line Selection,* P.J. Dix (ed.) VCH, Weinheim, pp. 257-286.

Gleddie, S., Keller, W. and Setterfield, G. (1983). Somatic embryogenesis and plant regeneration from leaf explants and cell suspensions of *Solanum melongena* (egg plant). *Can. J. Bot.,* **61:** 656-665.

Gleddie, S., Keller, W.A. and Setterfield, G. (1985). Plant regeneration from tissue, cell and protoplast cultures of several wild *Solanum* species. *J. Plant Physiol.,* **109:** 405-418.

Godwin, I., Todd, G., Ford-Lloyd, B. and Newburg, H.J. (1991). The effects of acetosyringone and pH on *Agrobacterium* mediated transformation vary according to plant species. *Plant Cell Rep.,* **9:** 671-675.

Gomez, O. and Chambonnet, D. (1992). *In vitro* androgenesis in Cuban F_1 hybrids of sweet pepper. *Capsicum Newsletter,* **11:** 26.

Gopalaswamienger, K.S. (1951). *Complete gardening in India.* The Hosali Press, Bangalore.

Gould, J., Banister, S., Hasegawa, O., Fahima, M. and Smith, R.H. (1991). Regeneration of *Gossypium hirsutum* and *G. barbadense* from shoot apex tissues for transformation. *Plant Cell Rep.,* **10:** 12-16.

Goyal, Y. and Arya, H.C. (1984). Effects of sugars, nitrogen, amino acids and vitamins on shoot differentiation from single bud *in vitro* culture of *Prosopis cineraria. Indian J. Exptl. Biol.,* **22:** 592-595.

Green, M.B. and Hedin, P.A. (1986). *Natural resistance of plants to pests: Role of allelochemicals. Amer. Chem. Soc.* Washignton DC (Acs. Symp. Ser. 296).

Greenway, H. and Munns, R. (1980). Mechanisms of salt tolerance in non-halophytes. *Ann. Rev. Plant Physiol.,* **31:** 149-190.

Grewal, S., Ahuja, A. and Atal, C.K. (1980). *In vitro* proliferation of shoot apices of *Eucalyptus citridera* Hook. *Indian J. Exp. Biol.,* **18:** 775-777.

Grover, A., Sanam, N. and Sahi, Ch. (1998a). Genetic engineering for high level tolerance of abiotic streses through over expression of transcription factor genes–The next frontier. *Curr. Sci.,* **75:** 178-179.

Grover, A., Pareek, A., Singla, S.L., Minhas, D., Katiyar, S., Ghawana, S., Dubey, H., Agarwal, M., Rao, G.U., Rathee, J. and Grover, A. (1998b). Engineering crops for tolerance against abiotic stresses through gene manipulation. *Curr. Sci.,* **75:** 689-696.

Grover, A., Kapoor, A., Lakshmi, O.S., Agarwal, S., Sahi, Ch., Agarwal, S.K., Agarwal, M. and Dubey, H. (2001). Understanding molecular alphabets of the plant abiotic stress response. *Curr. Sci.,* **80:** 206-216.

Guha, S. and Maheshwari, S.C. (1964). *In vitro* production of embryos from anther of *Datura*. *Nature,* **204:** 497.

Guha, S. and Maheshwari, S.C. (1966). Cell division and differentiation of embryos in the pollen grains of *Datura in vitro*. *Nature,* **212:** 97-98.

Guha, S. and Maheshawari, S.C. (1967). Development of embryoids from plllen grains of *Datura in vitro*. *Phytomorphology,* **17:** 454-461.

Gunay, A.L. and Rao, P.S. (1978). *In vitro* plant regeneration from hypocotyl and cotyledon explants of red pepper (*Capsicum*). *Plant Sci. Lett.,* **11:** 365-372.

Gupta, P.K., Nadgir, A.L., Mascarenhas, A.F. and Jagannathan, V. (1980). Tissue culure of Forest trees, clonal multiplication of *Tectona grandis* L. (Teak) by tissue culture. *Plant Science Lett.,* **17:** 259-268.

Gupta, S.K., Srivastava, A.K., Singh, P.K. and Tuli, R. (1997). *In vitro* proliferation of shoots and regeneration of cotton. *Plant Cell Tis. Org. Cult.,* **51:** 149-152.

Guri, A. and Sink, K.C. (1988). *Agrobacterium* transformation of egg plant. *J. Plant Physiol.,* **133:** 52-55.

Gu, S.R. (1979), Plantlets from isolated pollen culture of eggplant (*Solanum malongena L.). Acta Bot Sin., 21:* 30-36.

Haberlandt, G. (1902). Culture versuche mit isolierten Pflanzan Zellen. Math naturwiss K.L. Kais Akad. Wriss Wien, **111:** 69-92.

Hagemann, R. (1982). Induction of plastome mutations by nitroso urea compounds. In: *Methods in Chloroplast Molecular Biology,* (Eds.) Edelmann, M., Hallick, R.B. and Chau, N.H., pp. 119-127.

Hamilton, G.C., Jelenkovic, G.L., Lashomb, J.H., Ghidiu, G., Billings, S. and Patt, J.M. (1997). Effectiveness of transgenic eggplant (*Solanum malongena L)* against the Colorado potato beetle. *Adv. Hort, Soc. 11:* 189-192.

Handa, S., Handa, A.K., Hasegawa, P.M. and Bressan, R.A. (1986). Proline accumulation and the adaptation of cultured plant cells to water-stress. *Plant Physiology,* **80:** 938-945.

Hanning, E. (1904). Zur Physiologie Plfanzlicher Embryoness I. Uber die culture von Cruciferes Embryoness ausserhalb des Embryosacks. *Bot. Ztg.,* **62:** 45-80.

Hasan, E. and Konzak, C.E. (1997). Effects of light regimes on anther culture response in bread wheat. *Plant Cell Tiss. Org. Cult.,* **50:** 7-12.

Hayashi, H., Mustardy, A., Deshnium, P., Ida, M. and Murata, N. (1997). Transformation of *Arabidopsis thaliana* with Cod 'A' gene for choline oxidase accumulation of glycine-betaine and enhanced tolerance to salt and cold stress. *Plant,* **12:** 133-142.

Heinz, D.J., Krishnamurthy, M., Nickell, L.G. and Maretzki, A. (1977). Cell, tissue and organ culture in sugarcane improvement. In: *Applied and fundamental aspects of plant cell tissue and organ culture.* J. Reinert and Y.P.S. Bajaj (eds.) Springer Verlag, Berlin, pp. 3-17.

Heirwegh, K.M.G., Benerjee, N., Nerum, K. Van and Langhe E. (1985). Somatic embryogenesis and plant regeneration in *Cichorium intybus* L. (Witloof, Compositae). *Plant Cell Rep.,* **4:** 108-111.

Higgins, E., Hulme, J. and Shields, R. (1992). Early events in transformation of potato by *Agrobacterium tumefaciens. Plant Science,* **82:** 109-118.

Hinchee, M.A.W. (1998). Production of transgenic soybean plants using *Agrobacterium* mediated DNA transfer. *Biotechnology,* **6:** 915-922.

Hoekema, A., Huisman, M.J., Mosendyk, L., Vanden Elzen, P.J.M. and Cornelissen, B.J.C. (1989). The genetic engineering of two commercial potato cultivars for resistance of potato *virus x* Biol. *Biotechnology,* **7:** 273-278.

Hoftmann, E. (1970). *Steroid Biochemistry,* Acad. Press, New York.

Holmstrom, K.O., Mantyl, E., Wein, B. and Pelva, E.T. (1996). Drought tolerance in tobacco. *Nature,* **379:** 683-684.

Hood, E.E., Helmer, G.L., Fraley, R.T. and Chilton, M.D. (1986). The hypervirulence of *Agrobacterium tumefaciens* A 281 is encoded in a region of pTiBO 542 outside of T-DNA. *J. Bacteriol,* **168:** 1291-1301.

Hooykaas, P.J.J. and Schilperoort, R.A. (1992). *Agrobacterium* and plant genetic engineering. *Plant Mol. Biol.,* **13:** 327-336.

Hoque, A., Islam, R. and Arima, S. (2000). High frequency plant regeneration from cotyledon derived callus of *Mimordica dioica* (Roxb.) Willd. *Phytomorphology,* **50:** 267-272.

Hosoda, N. and Yatazava, M. (1979). *Agricult. and Biol. Chem.,* **43:** 821-825.

Hossain, M., Biswas, B.K., Karim, M.R., Rahman, S.M., Islam, R. and Joarder, O.I. (1994). *In vitro* organogenesis of elephant apple (*Feronia limonia*). *Plant Cell Tiss. Org. Cult.,* **39:** 265-268.

Hossain, M., Rahman, S.M., Zaman, A. and Joarder, O.I. (1991). Effect of nature of explants and pH on *in vitro* propagation of some mulberry genotypes. *Bull. Sericult. Res.,* **2:** 13-22.

Hosticka, L.P. and Hanson, M.R. (1984). Induction of plastid mutations in tomatos by nitrosomethyl urea. *J. Hered.,* **75:** 242-246.

Hsu, C.M., Yang, W.P., Chen, C.C., Lai, Y.K. and Lim, T.Y. (1993). A point mutation in the chloroplast rps, 12 gene from *Nicotiana* *plumbaginifolia* confers, Streptomycin resistance. *Plant Mol. Biol.*, **23:** 179-183.

Hu, C.Y. and Wang, P.J. (1983). Meristem, shoot tip and bud cultures. In: *Hand book of Plant Cell Culture Techniques of Propagation and breeding*, (eds.) Evans, D.A., Sharp, W.R., Ammirato, P.V. and Yamada Mcmillan, New York, **1:** 177-227.

Hu, H. and Zeng, J.Z. (1986). Development of new varieties via anther culture. In: *Hand book of plant cell culture*, Vol. 3, *Crop species* (eds.) P.V. Ammirato, D.A. Evans, W.R. Sharp and Y. yamada). Mc Millan Publishing Co. New York, pp. 65-90.

Hu, C.A.A., Delauney, A.J. and Verma, D.P.S. (1992). A bifunctional enzyme (Dpyrroline-5-carboxylate synthetase) catalyzes the first two steps in proline biosynthesis in plants. *Proceedings of the National Academy of Sciences*, USA, **89:** 9354-9358.

Hu, K.I., Motsubara, S. and Murakami, K. (1993). Haploid plant production by anther culture in carrot (*Daucus carota* L.). *J. Jpn. Soc. Hort. Sci.*, **62:** 561-565.

Huang, H.C. and Murashige, T. (1976). Plant tissue culture media, major constituents, their preparation and some applications. *Tissue Culture Association Manual*, **3:** 539-549.

Huettman, C.A. and Preece, J.E. (1993). Thiadiazuron: A potent cytokinin for woody plant tissue culture. *Plant Cell Tiss. Org. Cult.*, **33:** 105-119.

Huisman, M.J., Cornelissen, B.J.C. and Jongediyk, E. (1992). Transgenic potato plants resistant to viruses. *Euphytica*, **63:** 187-197.

Hussain, A., Jaiswal, U. and Jaiswal, V.S. (2000). Somatic embryogenesis and plantlet regeneration in Amrapali and Chausa cultivars of Mango (*Mangifera indica* L.). *Curr. Sci.*, **78:** 164-169.

Hussain, T., Malik, T., Raziuddin, S. and Gordon, M.P. (1997). Studies on the expression of marker genes in chick pea. *Plant Cell Tiss. Org. Cult.*, b49: **7-16.**

Hussain, M.M. 1995. Repellant effect of Katebegun (*Solanum xanthocarpum Sch.)* leaf on *Trilobium castaneium* Herbst, *Pakistan J. Zool, 27:* 279-280.

Ibrahim, K.M., Collins, J.C. and Collin, H.A. (1992). Characterization of progeny of *Coleus blumei* following an *in vitro* selection for salt tolerance. *Plant Cell Tiss. Org. Cult.,* **28:** 139-145.

Ingelbrecht, I., Breyne, P., Vancompernolle, K., Jacobs, A. Van Montagu, M., Depicker, A. (1991). Experimental analysis of transcriptional interference in transgenic plants. *Gene,* **109:** 239-242.

Isouard, G., Raguin, C. and Demarly, Y. (1979). Obtention de plantes haploids, et diploides per culture *in vitro,* anthers d aubergine (*Solanum melongena* L.). *C.R. Acad. Science Ser. D,* **288:** 987-989.

Jacobsen, H.J. and Kysely, W. (1984). Induction of somatic embryos in pea *Pisum sativum. Plant Cell Tiss. Org. Cult.,* **3:** 319-324.

Jacqueline, T.B. and Charlwood, V.B. (1986). The control of callus formation and differentiation in scented pelargoniums. *J. Plant Physiol.,* **123:** 409-417.

Jagadishchandra, K.S. and Satyanarayana, N. (1997). High frequency *in vitro* multiple shoot induction in mulberry var Mysore local (*Morus indica* L.). *J. Swamy Bot. Cl.,* **14:** 57-61.

Jagadishchandra, K.S., Rachappaji, S., Gowda, K.R.D. and Tharasaraswathi, K.J. (1999). *In vitro* propagation of *Pisonia alba* (L.) spanogae (Lettuce tree). A threatened species. *Phytomorphology,* **49:** 43-47.

Jain, S., Nainawatee, H.S., Jain, R.K. and Chowdhary, J.R. (1991a). Proline status of genetically stable salt tolerant, *Brassica juncea* L. somaclones and their parent. cv. Prakash, *Plant Cell Rep.,* **9:** 684-687.

Jain, R.K., Jain, S. and Chowdhary, J.B. (1991b). *In vitro* selection for salt tolerance in *Brassica Juncea* L. using cotyledon explants callus and cell suspension cultures. *Annals. Bot.,* **67:** 517-519.

Jain, R.K., Jain, S., Nainawatee, H.S. and Chowdhury, J.H. (1990). Salt tolerance in *Brassica juncea* L.–*in vitro* selection, agronomic evaluation and genetic stability. *Euphytica,* **48:** 141-152.

James, C. (1997). Global status of transgenic crops in 1997. ISAAA Briefs No. 5, ISAAA: Ithaca, New York, pp. 3.

Jarl, C.I., Ractveld, E.M. and Settaas, J.M. (1999). Transfer of fungal resistance through interspecific somatic hybridization between *Solanum melongena* and *S. torvum. Plant Cell Rep.,* **18:** 791-796.

JasRai, Y.T., Mudgil, Y., Remakanthan, A. and Kannan, V.R. (1999). Direct shoot regeneration from cultured leaves of *Passiflora caerulea* L. and field performance of regenerated plants. *Phytomorphology*, **49:** 289-293.

Jha, S. (1986). Production of polyploid plants through somatic embryogenesis from long term culture of diploid Indian Squill. *VI Int. Congr. Plant Tissue and Cell Culture, Minner poris,* Abst. 30.

Jia, S.R. and Chua, N.H. (1992). Somatic embryo-genesis and plant regeneration from immture embryo cultures of *Pharbitis nil. Plant Sci.*, **87:** 215-223.

Jones, J.D.G., Dean, C., Gidoni, D., Bond-Nutter, D., Lee, R., Bedrock, J. and Dunsmuir, P. (1988). Expression of bacterial chitinase protein in tobacco leaves using prophotosynthetic promoters. *Mol. Gen. Genet.*, **212:** 536-542.

Jones, J.D.G., Garland, F.M., Maliga, P. and Dooner, H.K. (1989). Visual detection of the maize element activator (AC) in tobacco. *Science*, **244:** 204-207.

Jones, J.D.G., Svab, Z., Harper, E.C., Hurueitz, C.D. and Maliga, P. (1987). A dominant nuclear streptomycin marker for plant cell transformation. *Mol. Gen. Genet.*, **210:** 86-89.

Jones, M. (1988). Fusing plant protoplasts. *Tib. Tech.*, **6:** 153-158.

Jordan, M., Pedraza, J. and Goreux, A. (1985). *In vitro* propagation studies of three prosopis species (*P. alba, P. chilensis* and *P. taumarugo*) through shoot tip culture. *Gartenbauwiseenschaft*, **50:** 265-267.

Kallack, H., Reidla, M., Hilpus, I. and Virumae, K. (1997). Effects of genotype, explant source and growth regulators on organogenesis in carnation callus. *Plant Cell Tiss. Org. Cult.*, **51:** 127-135.

Kalloo, G. (1993). Egg plant (*Solanum melongena* L.). In: *Genetic Improvement of Vegetable Crops.* Pergamon Press, pp. 587-604.

Kartha, K.K., Gamborg, O.L., Shyluk, J.P. and Constable, F. (1976). Morphogenetic investigations on *in vitro* leaf culture of tomato (*Lycoperis con esculentum* Mill. c.v. Starfire) and high frequency plant regeneration. *Z. Pflanzen Physiol.*, **77:** 292-301.

Kirtikar, K.V. and Basu, B.D. (2000). *Indian Medicinal Plants*, 9:2428-2430. Sri Satguru Publications, Delhi.

Kashyap, V., Douval, A., Verma, A. and Rajam, M.V. (1999). Plant regeneration in wild species of egg plant. In: *Proc. Nat. Symp. on Role of Plant Tissue Culture in Biodiversity Conservation and Economic Development*, Almora, India, pp. 14-15,.

Kashyap, V., Vinod Kumar, S., Collonnier, C., Fusari, F., Haicour, R., Rotino, G.L., Sihachakr, D. and Rajam, M.V. (2003). Biotechnology of egg plant. *Scientia Horticulturae, 97:* 1-25.

Kathiravan, K. and Ignacimuthu, S. (1999). Micropropagation of *Canavalia virosa* (Roxb.). *Phytomorphology.,* **49:** 61-66.

Kaur, R., Mahajan, R., Bhardwaj, S.V. and Sharma, D.R. (2000). Meristem tip culture to eliminate strain berry mottle virus from two cultivars of straw berry. *Phytomorphology,* **50:** 192-194.

Kavi Kishor, P.B., Hong, Z., Miao, G.H., Hu, C.A. and Varma, D.P.S. (1995). Over expression of pyrroline-5-carboxylate synthetase increases proline production and confers osmotolerance in transgenic plants. *Plant Physiol.,* **108:** 387-1397.

Keller, W.A. *et al.* (1987). In: *Plant Tissue and Cell Culture Recent Developments and Future Prospects* (eds.), Green, C.E. *et al.*, A.R. Liss, pp. 223-241.

Kemper, E., Grevelding, Ch., Schell, J. and Masterson, R. (1992). Improved method for the transformation of *Arabidopsis thaliana* with chimeric dihydrofolate reductase constructs which confer methotrexate resistance. *Plant Cell Rep.,* **11:** 118-121.

Khehra, G.S. and Mathias, R.J. (1992). The interaction of genotype, explant and media, on the regeneration of shoots from complex explants of *Brassica rapus* L. *J. Exp. Bot.,* **43:** 1413-1418.

King, L.S. and Rao, A.N. (1981). Induction of callus and organogenesis In: *Cocoa tissues.* In *Pros. Costed Symp. on Tiss. Cult. Of Economically important plants.* pp. 107-112.

King, P.J. (1984). Mutagenesis of cultured cells. In: *Cell Culture and Somatic cell genetic of plants, Vol. 1,* (ed.) Vasil IK, Academic Press, New York, pp. 547-551.

Kirk, J.T.O. and Tilney-Basset, R.E.R. (1978). *The plastids, their chemistry, structure, growth and inheritance,* Elsevier, North Holland, Amsterdam.

Kitamura, Y. (1988). *Duboisa* spp: *In vitro* regeneration and the production of tropane and pyridine alkaloids. In: *Biotechnology in Agriculture and Forestry-Medicinal and Aromatic Plants*. Bajaj YPS (ed.), **1:** 419-436.

Knapp, S.J. (1991). Using molecular markers to map multiple quantitative trait loci models for back cross, recombinant inbred doubled haploid progeny. *Theor. Appl. Genet.*, **81:** 333-338.

Kobayashi, R.S., Stomme, J.R. and Sinden, S.I. (1996). Somatic hybridization between *Solanum ochiranthum* and *Lycopersicon esculentum*. *Plant Cell Tiss. Org. Cult.*, **45:** 73-78.

Konar, R.N. and Kitchlue, S. (1982). In: *Experimental Embryology of Vascular plants*, (ed.). B.M. Johra, pp. 53-78.

Kowzyk, T.P., Mackenzie, I.A. and Cocking, E.C. (1983). Plant regeneration from organ explants and protoplasts of medicinal plant *Solanum khasianum* C.B. Clarke var. Chatter Geanum Sneguta (Syn. *Solanum viarum* Dundl.). *Z. Pflanzen. Physiol.*, **11:** 55-68.

Krishnamurthy (1982). In: *Tissue culture of economically important plants*. (ed.) Rao, A.N., Costed and Singapore, pp. 70.

Kristiansen, K. and Andersen (1993). Effect of donor plant, temperature, photoperiod and age on anther culture response of *Capsicum annum* L. *Euphytica*, **67:** 105-110.

Kukreja, A.K. and Mathur, A.K. (1985). Tissue culture studies in *Duboisia myoporoides* L. Plant regeneration and clonal propagation by stem node culture. *Planta Med.*, **2:** 93-96.

Kumar, P.A., Mandaokar, A. Sreenivasu, K., Chakrabarthi, S.K., Bisaria, S., Sharma, S.R., Kaur, S. and Sharma, R.P. (1998). Insect resistant transgenic brinjal plants. *Mol. Breed.*, *4:* 33-37.

Kumari, A.S. and Kumar, A. (1995). Plant regeneration from cultured embryogenic axis *Thevetia peruviana* L. *Indian J. Exptl. Biol.*, **33:** 190-193.

Kumari, N. Jaiswal, U, and, Jaiswal, V.S. (1998). Induction of somatic embryogenesis and plant regeneration from leaf callus of *Terminalia arjuna*. *Curr. Sci.*, **25:** 1052-1055.

Laibach, F. (1925). Das Taubverden von Bastardsmen and die Kunsliche Aufzucht fruh baster bender Bastardem bryonen. *Z. Bot.*, **17:** 417-459.

Laibach, F. (1929). Ectogenesis in plants, method and genetic possibilities of propagating embyros otherwise dying in the seed. *J. Hered,* **20:** 201-208.

Lakshmana Rao, P.V. and Singh, B. (1991). Plantlet regeneration from encapsulated somatic emrbyos of hybrid *Solanum melongena* L. *Plant Cell Rep.,* **10:** 7-11.

Lakshmi Sita, G. and Chatopadhyay, S. (1986). In: *Plant Cell Tissue Culture of Economically important plants,* Reddy, G.M. (Ed.) 195.

Lakshmi Sita, G. and Vaidyanathan, C.S. (1979). Rapid multiplication of *Eucalyptus* by multipleshoot production. *Curr. Sci.,* **48:** 350-351.

Lakshmisita, G., Raghava Ram, N.V. and Vaidyanathan, C.S. (1980). "Triploid plants from endosperm cultures of sandalwood by experimental embryogenesis". *Plant Sci. Lett.,* **20:** 63-69.

Lal, N. and Ahuja, P.S. (1989). Propagation of Indian Rhubarb (*Rheum emodi* Walt) using shoot tip and leaf explant culture. *Plant Cell Rep.,* **8:** 493-496.

Lansmier, E.M. and Skoog, F. (1965). Organic growth factor requirement for tobacco tissue cultures. *Physiol. Plant.,* **18:** 100-127.

Larkin, P.J. and Scrowcroft, W.R. (1981). Somaclonal variation–A novel source of variability from cell cultures for plant improvement. *Theor. Appl. Genet.,* **60:** 197-214.

Lawrence, P.K. and Koundal, K.R. (2001). *Agrobacterium tumefaciens* mediated transformation of pigeon pea (*Cajanus cajan* L. Mill. Sp.) and molecular analysis of regenerated plants. *Curr. Sci.,* **80:** 1428-1432.

Lee, J.H., Hubel, A. and Schoffl, F. (1995). Depression of the activity of genetically engineered heat shock factor causes constitutive synthesis of heat shock proteins and increased thermotolerance in transgenic *Arabidopsis. Plant J.,* **8:** 603-612.

Lee, J.H., Montagu, M.V. and Verbruggen, N. (1999). A highly conserved kinase is an essential component for stress tolerance in yeast and in plant cells. *Proc. Natl. Acad. Sci. (USA),* **96:** 5873-5877.

Leone, M., Filippone, E. and Lurquin, P.F. (1993). Transformation in *Solanum melongena* L. (egg plant). In: Bajaj (ed.) *Biotechnology in*

Agriculture and Forestry: Plant Protoplasts and Genetic Engineering (pp. 320-328), Springer, Verlag-Berlin, Germany.

Leone, A., Costa, A., Tucci, M. and Grillo, S. (1994). Adaptation, versus shock response to polyethylene glycol induced low water potential in cultured potato cells. *Physiol. Plant.*, **92:** 21-30.

Letham, D.S. (1968). A new cytokinin bioassay and naturally occurring cytokinin complex. In: *Biochemistry and Physiology of Plant hormones.* (eds.) F. Wightman and Selterfield.

Lindsey, K. and Yeoman, M.M. (1983). Novel experimental system for studying the production of secondary metabolites by plant tissue culture. In: *Plant Biotechnology,* (eds.) S.H. Maritell and H. Smith, Cambridge University, Cambridge, pp. 39-66.

Litz, R.E., Knight, R.J. and Gazit, S. (1982). Somatic embryos from cultured ovules of polyembryonic, *Mangifera indica* L. *Plant Cell Rep.*, **1:** 264-266.

Litz, R.E., Knight, R.J. and Gazit, S. (1983). *In vitro* somatic embryogenesis from *Mangifera indica* L. Callus. *Science Hortic.*, **22:** 233-240.

Liu, W., Parrott, W.A., Hildebrand, D.F., Collins, G.B. and Williams, E.G. (1990). *Agrobacterium* induced gall formation in bell pepper (*Capsicum annum* L.) and formation of shoot-like structures expressing introduced genes. *Plant Cell Rep.*, **9:** 360-364.

Lorze, H., Goebel, E. and Brown, P. (1988). Advances in tissue culture and progress towards genetic transformation of cereals. *Plant Breed.*, **106:** 1-25.

Magioli, C., Rocha, A.P.M.,de Oliveria, D.E. and Mansus, E. (1998). Efficient shoot organogenesis of egg plant (*Solanum melongena* L.) induced by thidiazuron. *Plant Cell Rep.*, **17:** 161-163.

Maheswari, S.C., Tyagi, A.K., Malhotra, K. and Sopory, S.K. (1980). Induction of haploids from pollen grains in angiosperms–The Current Status. *Theor. Appl. Genet.*, **58:** 193-206.

Malathy, S. and Pai, J.S. (1998). Micropropagation of *Ixora singaporensis* (Linn.)–An ornamental Shurb. *Curr. Sci.*, **75:** 545-547.

Maliga, P. (1980). Isolation, characterization and utilization of mutant cell lines in higher plants. *Int. Rev. Cytol. Suppl.*, **11A:** 225-250.

Maliga, P. (1981). Streptomycin resistance is inherited as a recessive mendelian trait in *Nicotiana sylvestris* Line. *Theor. Appl. Genet.*, **60:** 1-3.

Maliga, P. (1984). Isolation and characterization of mutants in plant cell culture. *Ann. Rev. Plant Physiology*, **35:** 519-542.

Maliga, P., Svab, Z., Harper, E.C.and Jones, J.D.G. (1988). Improved, expression of streptomycin resistance in plant due to a deletion in the streptomycin phosphotransferase coding sequence. *Mol. Gen. Genet.*, **214:** 456-459.

Mandal, N. and Gupta, S. (1997). Anther culture of an inter specific rice hybrid and selection of fine grain type with submergence tolerance. *Plant Cell Tiss. Org. Cult.*, **51:** 79-82.

Mandal, A., Chowdhary, B. and Sheeja, T.E. (2000). Development and characterization of salt tolerant somaclones, in rice cultivar pakkali. *Indian J. Exptl. Biol.*, **38:**.

Manske, H.F. and Halmes, H.L. (1970). *The alkaloids: Chemistry and Physiology*, Academic Press Inc., New York.

Mantell, S.H., Mathews, J.A. and McKee, R.A. (1985). *Principles of Plant Biotechnology*, Blackwell Scientific Publications, London.

Maria, T.H., Margarito, C.M.P. and Tarrogo, J.F. (1990). One step shoot tip multiplication and rooting of *Digitalis thapsil*. *Plant Cell Tiss. Org. Cult.*, **22:** 179-182.

Mariani, P. (1992). Egg plant somatic embryogenesis combined with synthetic seed technology. In: *Proc. 8th meeting on Genetics and Breeding of Capsium and Eggplant*, Rome, Italy, pp. 289-294.

Martha, C.W., Sandra, M.R., Joyce, A.B. and Wynne, J.C. (1991). Effect of microspore stage and media on anther culture of peanut (*Arachis hypogaea* L.). *Plant Cell Tiss. Org. Cult.*, **24:** 25-28.

Mascarenhas, A.F. and Muralidharan, E.M. (1989). Tissue culture of forest trees in India. *Curr. Sci.*, **58:** 606-613.

Mathur, A., Mathur, A.K., Kukreja, A.K., Ahuja, P.S. and Tyagi, B.R. (1987). Establishment and multiplication of colchi autotetraploids of *Rauvolfia serpentina* L. *Plant Cell Tiss. Org. Cult.*, **10:** 129-134.

Mathur, J., Ahuja, P.S., Mathur, A., Kukraya, A.K. and Shah, N.C. (1988). *In vitro* propagation of *Valeriana wallichi*. *Plant Med.*, **54:** 82-83.

Matsuoka, H. and Hinata, K. (1979). NAA-induced organogenesis and embyogenesis in hypocotyl callus of *Solanum melongena* L. *J. Exptl. Bot.,* **30:** 363-370.

Miyoshi, K. (1996). Callus induction and plantlet formation through culture of isolated microspores of egg plant (*Solanum malongena L.*). *Plant Cell Rep. 15*:391-395.

McCabe, P.F., Timmons, A.M. and Dix, P.J. (1989). A simple procedure for the isolation of streptomycin resistant plants in *Solanaceae. Mol. Gen. Genet.,* **216:** 132-137.

McCabe, D.E. and Martinell, B.J. (1993). Transformation of elite cotton cultivars via particle bombardment of meristems. *Biotechnology,* **11:** 596-598.

McCormick, S., Nicadarmeyer, J., Fty, J., Barnason, A., Hossb, R., Fraley, R. (1986). Leaf disk transformation of cultivated tomato (*L. esculentum*) using *Agrobacterium tumefaciens. Plant Cell Rep.,* **5:** 81-84.

McCormick, S. (1991). In: *Plant Tissue Culture Manual Fundamentals and Applications,* (ed.) Lindsley, K., Kulwer Acad. Publ., Dordrecht, The Netherlands, pp. B6 1-9.

McCoy, T.J. (1987). Characterization of alfalfa (*Medicago sativa* L.). Plants regenerated from selected NaCl tolerance cell lines. *Plant Cell Rep.,* **6:** 417-422.

McGranahan, G.H., Charles, A.L., Uratsu, S.L. and Dandekar, A.M. (1989). Improved efficiency of the walnut somatic embryos gene transfer system. *Plant Cell Rep.,* **8:** 512-516.

McHugen, A. (1987). Salt tolerance through increased vigour in a flax line (STS-11) selected for salt tolerance *in vitro. Theor. Appl. Genet.,* **74:** 722-732.

Medgyesy, P., Menczel, L. and Maliga, P. (1980). The use of cytoplasmic streptomycin resistance chloroplast transfter *Nicotiana tabacum* into *Nicotiana sylvestris* and Isolation of their somatic hybrids. *Mol. Gen. Genet.,* **179:** 693-698.

Medgyesy, P. (1990). Selection and analysis of cytoplasmic hybrids. In: *Plant Cell Line Selection,* P.J. Dix (ed.), Velt, Weinheim, pp. 287-316.

Mehra-Palta, A. (1982). Clonal propagation of *Eucalyptus* by Tissue Culture. *Plant Sci. Lett.,* **26:** 1-11.

Meins, F. (1983). Heritable variation in plant cell culture. *Ann. Rev. Plant Physiol.,* **34:** 327-346.

Menczel, L., Polsby, L.S., Steinback, K.E. and Maliga, P. (1986). Fusion mediated transfer of triazine-resistant chloroplasts: characterization of *Nicotiana tabacum* cybrid plants. *Mol. Gen. Genet.,* **205:** 201-205.

Merceir, H., Vieira, C.C.J. and Figueredo-Ribeiro, R.C.L. (1992). Tissue culture and plant propagation of *Gomphrena officinalis* a Brazilian medicinal plant. *Plant Cell Tiss. Org. Cult.,* **28:** 249-254.

Meyer, P., Kartzke, S., Niedenhof, I., Heidmann, I., Bussman, K.and Saedler, H. (1988). A genomic DNA segment from *Petunia hybrida* leads to increased transformation frequencies and simple integration patterns. *Proc. Natl. Acad. Sci.* (USA), **85:** 8568-8572.

Mhatre, M.M., Bapat, V.A. and Rao, P.S. (1984). Plant regeneration in protoplast cultures of *Tylophora indica. J. Plant Physiol.,* **115:** 231-235.

Mitchell, M.J., Busch, R.H. and Rines, H.W. (1992). Comparison of lines derived by anther culture and single-seed descent in a spring wheat cross. *Crop. Sci.,* **32:** 1446-1451.

Mitgko, J., Andrasfalvy, A., Csillery, G. and Fari, M. (1995). Anther culture response in different genotypes and F_1 hybrids of pepper (*Capsicum annum* L.). *Plant Breeding,* **114:** 78-80.

Mitra, D.K. and Gupta, N. (1989). Somaclonal variant for tolerance to little leaf disease of egg plant (*Solanum melongena* L.) regenerated from tissue culture of infected plants. *New Bot.,* **16:** 291-292.

Mitra, D.K., Gupta, N. and Bhaskaran, S. (1981). Development of plantlets from brinjal (*Solanum melongena* L.) stem tissues infected with little leaf, a mycoplasma disease. *Indian J. Exptl. Bot.,* **19:** 1177-1178.

Mochizuki, H. and Yamakawa, K. (1979). Resistance of selected egg plant cultivars and related wild *Solanum* sps. to bacterial wilt (*Pseudomonas solanacearum*). *Bull. Veg. and Ornamental Crops Res.,* **6:** 1-10.

Mohan, J.S.S., Vijayakumar, V., Aparna, V. and Vaidya, R.P. (2000). Somatic embryogenesis and plant regeneration in *Tribulus terrestris* (L.). *Phytomorphology,* **50:** 307-311.

Moieni, A. and Sarrafi, A. (1995). Genetic analysis for haploid regeneration responses of hexaploid-wheat anther cultures. *Plant Breeding,* **114:** 247-249.

Morrison, R.A. and Evans, D.A. (1988). Haploid plant from tissue culture: new plant variations in a shortened time frame. *Biotechnology,* **6:** 684-690.

Muir, W.H. (1953). *Cultural conditions favouring the isolation and growth of single cell from higher plants in vitro,* Ph.D. thesis, University of Wisconsin, U.S.A.

Mujib, A., Bandyopadhyay, S., Banerjee, S. and Ghosh, P.D. (1990). Somatic embryogenesis from, hypocotyl callus of *Coriander* (*Coriandrum sativum* L.). *The Hort. J.,* **3:** 59-62.

Mujib, A., Bandyopadhyay, S., Jana, B.K. and Ghosh, P.D. (1998). Direct somatic embyogenesis and *in vitro* plant regeneration in *Hippeastrum hybridum. Plant Tissue Cult.,* **8(1):** 19-25.

Murashige, T. and Skoog, F. (1962). A revised medium for rapid growth and bioassay with tobacco tissue culture. *Physiol. Plant.,* **159:** 473-497.

Murashige, T. (1974). Plant propagation through tissue culture. *Ann. Rev. Plant Physiol.,* **23:** 135-166.

Murukami, Y., Tsuyama, M., Kobayashi, Y., Kodama, H. and Iba, K. (2000). Trienoic fatty acids and plants tolerance of high temperature. *Science,* **287:** 476-479.

Nadkarni, A.K. (1954). *Indian Materia Medica-I,* Popular Book Depot, Bombay, pp. 1156.

Narayanaswamy, S. (1977). Regeneration of plants from tissue cultures. In: *Applied and fundamental aspects of plant cell, tissue and organ culture.* eds., J. Reinert and Y.P.S. Bajaj, Springer-Verlag, Berlin, pp. 179-206,

Nasir, A., Saeed, Yusuf Zafer and Kauser, A. Malik. (1997). A simple procedure of *Gossypium* meristem shoot tip culture. *Plant Cell Tiss. Org. Cult.,* **51:** 201-207.

Nayar, M.P., Ramamurthy, K. and Agarwal, V.S. (1989). *Economic plants of India*, Vol. I to V, Published by Director B.S.I., Calcutta.

Nazeem, P.A., Raji, P., Sherly, S., Lissamma, J. and Nybe, E.V. (1997). Indian Society for Spices, In: *Biotechnology of spices, medicinal and aromatic crops*. (eds.) Edison, S., Ramana, K.V., Sasikumar, B., Nirmal Babu, K. and Santhosh, J.E., pp. 87-93.

Neeta, D.S., Leela, G. and Susan, E. (2000). Direct regeneration of shoots from immature inflorescence cultures of turmeric. *Plant Cell Tiss. Org. Cult.*, **62:** 235-238.

Neeta, D. S., Singh, H., Tivarekar, S. and Eapen, S. (2001). Plant regeneration from different explants of neem. *Plant Cell Tiss. Org. Cult.*, **65:** 159-162.

Negruitiy, I., Jacobs, M. and Caboche, M. (1984). Advances in somatic cell genetics of higher plants–The protoplast approach in basic studies on mutagenesis and isolation of biochemical mutants. *Theor. Appl. Genet.*, **67:** 289-304.

Nesslar, C.L. (1982). Somatic embryogenesis in the opium poppy *Papaver somniferum*. *Physiol. Plant*, **55:** 453-458.

Niranjan, M.H. and Sudarshana, M.S. (2000). *In vitro* plant regeneration in *Nymphoides cristatum* (Roxb.) O. Kuntze. *Phytomorphology*, **50:** 343-344.

Nitsch, J.P. and Nitsch, C. (1969). Haploid plants from pollen grains. *Science*, **163:** 85-87.

Nobecourt, P. (1939). Cultures en sterie de tissus vegetaux sur milieu artificiel, *C.R. Seanc, Soc. Biol.*, **205:** 521-523.

Novak, F.J. and Havel, L. (1981). Shoot production from *in vitro* cultured flower heads of *Allium porrum* L. *Biol. Plant*, **23:** 266-269.

Noyak, F.J. and Mickel, A. (1987). *In vitro* mutants technology for crop improvement in developing countries, *Sabroa*, 19.

Omar, M.S. (1988). *Rhazya stricta* Decaisne: *In vitro* culture and the production of indole alkaloids. In: *Biotechnology in Agriculture and Forestry*, Bajaj, Y.P.S. (ed.), Springer, Berlin, Heidelberg, New York, Tokyo, pp. 529-540.

Ooms, G., Bains, A., Burrel, M., Karp, A., Twel, D. and Wilcox, E. (1987). Genetic manipulations in cultivars of oil seed rape

Brassica napus using *Agrobacterium*. *Thero. Appl. Genet.*, **71:** 325-329.

Paek, K.Y., Chandler, S.F. and Thorpe, T.A. (1988). Physiological effects of Na_2SO_4 and NaCl on callus cultures of *Brassica campestris* (Chinese cabbage). *Physiol. Plant.*, **72:** 160-166.

Pang, S.Z. and Sanford, J.C. (1988). *Agrobacterium* mediated gene transfer in papaya. *J. Amer. Soc. Hort. Sci.*, **113:** 287-291.

Parimerd, C., Vedel, F., Mathieu, C., Pelletier, G. and Chevre, A.M. (1988). Interspecific somatic hybridization between *Brassica napus* and *Brassica nirta* (*Sinapis alba* L.). *Theor. Appl. Genet.*, **75:** 546-552.

Parrott, W.A., Merkle, S.A. and Williams, *E.G.* (1991). Somatic embryogenesis potential for use in propagation and gene transfer systems. In: *Advanced methods in Plant Breeding and Biotechnology*, (ed.) D.R. Murrey, CAB International, Willing Ford Oxon (U.K.), pp. 158-200.

Parthasarathy, V.A., Bose, T.K., Deka, P.C. (2001). *Biotechnology of Horticultural Crops*, vol. 2 and 3, Naya Prakash Publishers, Calcutta.

Pattanaik, S.K., Sahoo, Y. and Chand, P.K. (1995). Efficient plant retrieved from alginate encapsulated vegetative buds of mature mulberry trees. *Scientia Hort.*, **49:** 1-13.

Pattnaik, S.K. and Chand, P.K. (1996). *In vitro* propagation of medicinal herb *Ocimum americanum* L. syn O. *canum* Sims (holy absil) and *Ocimum sanctum* L. (holy basil). *Plant Cell Rep.*, **15:** 846-850.

Pawar, P.K., Pawar, C.S., Narkhede, B.A., Teli, N.P., Bhalsingh, S.R. and Maheshwari, V.L. (2002). A technique for rapid micropropagation of *Solanum surattense* Burm. f. *Indian J. Biotech.*, **1:** 201-204.

Perez-Alfocea, F., Guerrier, G., Estah, M.I. and Bolarin, M.C. (1994). Comparative salt response at cell and molecular plant levels of cultivated and wild tomato species. *J. Hort. Sci.*, **69:** 639-644.

Philip, S., Banerjee, N.S. and Das, M.R. (2000). Genetic variation and micropropagation in three varieties of *Piper longum* L. *Curr. Sci.*, **78:** 169-173.

Philips, G.C. and Hubstenberger, J.F. (1985). Organogenesis in pepper tissue culture. *Plant Cell Tiss. Org. Cult.,* **4:** 262-269.

Pierik, R.L.M. (1987). *In vitro* culture of higher plants. Nartinus Mijhoft Dordrecht, pp. 183-230.

Pike, L.M. and Yas, K.S. (1990). A tissue culture tecnique for the clonal propagation of onion using immature flower buds. *Scientia Horti.,* **45:** 31-36.

Potrykus, I. (1990). Gene transfer to cereals: An assessment. *Biotechnology,* **8:** 535-542.

Prabhavathi, V., Yadav, J.S., Kumar, P.A. and Rajam, M.V. (2002). Antibiotic stress tolerance in transgenic egg plant (*Solanum malongena L.)* by introduction of bacterial mannitol phosphodeh drogenase gene. *Mol. Breed, 9*: 137-147.

Prasad, R.N. and Chaturvedi, H.C. (1978). *In vitro* induction of shoots and formation of plantlets from segments of leaf, stem and root of *Solanum xanthocarpum* Schrad. and Klendl. *Indian J. Exptl. Biol.,* **16:** 1121-1122.

Prasad, R.N., Sharma, M., Sharma, A.K. and Chaturvedi, H.C. (1998). Androgenic stable somaclonal variant of *Solanum surattense* Burm. f. *Indian J. Exptl. Biol.,* **36**: 1007-1012.

Praveen, M., Lakshman, A., Ugandhar, T. and Ramaswamy, N. (2001). Callusing efficiency and plant regeneration from different explants of *Strychnos potatorum*–A medicinally important forest tree. In: *Proc. On Frontiers of Plant Biotechnology* (eds.) Sadanandam, A., Reddy, K.J.M., Reddy, S.Ram Reddy and Rama Swamy, N., pp. 183-197.

Price, H. and Smith, R. (1984). *Hand book of plant cell cultures,* (eds). Ammirato, P.V., Evans, S., W. and Yamada, Y., **3:** 487-510

Purohit, S.D., Dave, A. and Kukda, G. (1994). Micropropagation of safed musli (*Chlorophytum borivillianum*) a rare medicinal herb. *Plant Cell Tiss. Org. Cult.,* **39:** 93-96.

Qin, X. and Rotino, G.L. (1993). Anther culture of several sweet and hot pepper genotypes. *Capsicum* and *Egg plant News Letter,* **12:** 59-62.

Quak, F. (1977). Meristem culture and virus-free plants. In: *Applied and fundamental aspects of plant cell, Tissue and Organ Culture,*

(Eds.) Reinert, J. and Bajaj, Y.P.S. Springer-Verlag, Berlin, pp. 598-615.

Raghavan, V. (1977). Applied aspects of embryo culture. In: *Applied and fundamental aspects of plant cell tisue and organ culture*, J. Reinert and Y.P.S. Bajaj (eds.), Springer-Verlag, Berlin, pp. 375-397.

Raghawan, V. (1986). Variability through wide crosses and embryo rescue. In: *Cell Culture and Somatic Cell Genetics of Plants*, Vol. 3. Plant Regeneration and Genetic Variability. (Ed.), I.K. Vasil, Academic Press, Orlando, pp. 613-633.

Raghavan, V. (1994). *In vitro* methods for the control of fertilization and embryo development. In: *Plant Cell and Tissue Culture*, Vasil, I.K. and Thorpe, T.A. (eds.). Kluwer Acad. Pub., Dordrecht, Netherlands, pp. 173-194.

Raguraman, G. and Ramanujam, M.P. (1998). Micropropagation of *Melia azaderach* (L.). *J. Swamy Bot. Cl.*, **15:** 1-5.

Rains, D.W. (1989). Plant tissue and protoplast culture, Applications to stress physiology and biochemistry. In: *Plants under stress.* (eds.) Jones, H.G., Flowers, T.J. and Jones, M.B., *Society for Experimental Biology Seminar Series*, **39:** 181-196, Comb University Press.

Rajasekharan, K., Mullins, M. and Nair, Y. (1983). Flower formation *in vitro* by hypocotyl explants of Cucumber (*Cucumis sativus*). *Annals. Bot.*, **52:** 417-420.

Rama Swamy, N., Hille, J., Van Haaren, J.J., Mark Tarcies Kneppers, Maart, J.I. and Nijkamp, H. John (2000). Genetic transformation studies on tomato. *Rec. Tr. Pt. Sci.*, **83:** 100.

Rama Swamy, N., Hille, J., Van Haaren, J.J. Mark, Tarcies Kneppers and Nijkamp, H. John (2006d). Efficient protocol for *Agrobacterium* mediated genetic transformation tobacco. *Plant Cell Reports.*

Rama Swamy, N., Ugandhar, T., Praveen, M. Lakshman, A., Rambabu, M., and Venkataiah, P. (2004). *In vitro* propagation of medicinally important *Solanum surattense, Phytomorphology 54:* 281-289.

Rama Swamy, N., Ugandhar, T., Praveen, M., Rambabu, M., and Upender, M., (2005a). Induction of streptomycin–resistant plantlets in *Solanum surattense* through *in vitro* mutagenesis. *Plant Cell Tiss. and Org. Cult., 80:* 201-207.

Rama Swamy, N., Ugandhar, T., Praveen, M., Venkataiah, P., Rambabu, M., Upender, M. and Subhash, K. (2005b). Somatic embryogenesis and Plantlet regeneration from cotyledon and leaf explants of *Solanum surattense. Indian J.Biotech,* **4:** 414-418.

Rama Swamy, N., Ugandhar, T., Praveen, M., Upender, M. and Subhash, K. (2006a). *In vitro* regeneration from different explants in *Solanum surattense. Trends in Biotech.* (In Press).

Rama Swamy, N., Ugandhar, T., Upender, M., Rambabu, M. and Praveen, M. (2006b). Nacl/Kcl/Mannitol tolerance cell line selection in *Solanum surattense. J. Genet. & Br.* (accepted).

Rama Swamy, N., Ugandhar, T., Praveen, M., Upender, M. and Subhash, K. (2006c). Efficient plant regeneration in *Solanum surattense. Indian J. Genet. & Br.* (In Press).

Rama Swamy, N., Ugandhar, T., Praveen, M., Rambabu, M. and Upender, M. (2006b). Micropropagation of *Solanum surattense. In vitro Biol. Plant* (accepted).

Rao, A.V., Venu, Ch. And Sadanandam, A. (1997). Selection of streptomycin and kanamycin resistance using nitrosomethyl-urea and *Agrobacterium* in *Solanum sisymbrifolium. Indian J. Exptl. Biol.,* **35:** 188-192.

Rao, A.V., Farooqui, A., Jayasree, T., Ramana, R.V. and Sadanandam, A. (1993). EMS-induced streptomycin resistance in *Solanum melongena* L. *Theor. Appl. Genet.,* **87:** 527-530.

Rao, G.P., Reddy, K.R.K. and Bir Bahudur (1989). *In vitro* morphogenesis from hypocotyl and cotyledonary callus cultures of *Turnera subulata.* J.E. Smith (Turneraceae). *Adv. Plant Sci.,* **2:** 100-105.

Rao, M.M. and Lakshmi Sita, G. (1996). Direct somatic embryogenesis from immature embryos of rosewood (*Dalbergia latifolia* Roxb.). *Plant Cell Rep.,* **15:** 355-359.

Rao, P.S. and Harada, H. (1974). Hormonal regulation of morphogenesis in organ culture of *Petunia inflata, Antirhinum majas* and *Pharbitis nil.* In: *Plant Growth Substances,* Hirokwa Tokyo, pp. 1113-1120.

Rao, P.S. and Swamy, S. (1972). Morphogenetic investigations in callus cultures of *Tylophora indica. Physiol. Plantarum.,* **27:** 271-276.

Rashid, A. (1988). *Cell physiology and genetics of higher plants,* (ed.) Boca Raton, F.L., CRC Press, Vol. I., pp. 1-38, 67-103.

Rauber, M. and Groune Waldt, J. (1988). *In vitro* regeneration in *Allium* species. *Plant Cell Rep.,* **7:** 426-429.

Ravindra Sharma (2003). *Medicinal Plants of India An Encyclopaedia,* Daya publishing House, Delhi.

Razdan, M.K. and Cocking, E.C. (1981). Improvement of legumes by explaining extra specific genetic variation. *Euphytica,* **30:** 819-833.

Rech, E.L. and Pires, J.P. (1986). Tissue culture propagation of *Mentha* spp. by the use of axillary buds. *Plant Cell Rep.,* **5:** 17-18.

Reddy, L.R. and Reddy, G.M. (1993). Factors affecting direct somatic embryogenesis and plant regeneration in groundnut (*Arachis hypogaea* L.). *Indian J. Exptl. Biol.,* **31:** 57-60.

Redenbaugh, K., *et al.* (1993). Regulatory issues for commercialization of tomatoes with an antisense polygalacturonase gene. *In vitro Cell Dev. Biol.,* **29:** 17-26.

Redenbaugh, K., Fuju, J. and Slade, D. (1991). Synthetic seed technology. In: *Cell Culture and Somatic Cell Genetics of Plants, Vol. 8, Scaleup and Automation in Plant Tissue Culture,* Vasil, J.K. (ed.), Academic Press, Orlando, pp. 35-74.

Reed, A.J., Magin, K.M., Anderson, J.S., Austin, G.D., Rangwala, T., Linde, D.C. Love, J.N., Rogers, S.G. and Fuchs, R.L. (1995). *J. Agri. Fol. Chem.,* **43:** 1954-1962.

Reinert, J. (1958). Morphogenese and ihre, kontrolle an gewebekulturen aus carotten. *Natusuissen schaften,* **45:** 344-345.

Reinert, J. (1959). Uber die Kontrolle der morphogenese and die induktion von adventive embryonen an Gewebekuluren aus aus Kartten. *Planta,* **58:** 318-333.

Roberts, A.V., Yokoya, K., Walker, S. and Mottley, J. (1995). Somatic embryogenesis in woody plants. (eds) Jain, S., Gupta, P. and Newton, R. Kluwer Academic Publishers, The Netherlands, 277-289.

Roddick, J.G., Riynenberg, A.L. and Wiessenberg, M. (1992). Alterations to the permeability of liposome membranes by the solasodine-based glycoalkaloids Solasonine and Solamargine. *Phytochem.*, **31:** 1951-1954.

Roest and Gilisen, L.J.W. (1989). Plant regeneration from protoplast– A literature review. *Acta Bot. Neerl.*, **38:** 1-23.

Romeij, G. and Van Lammeren, A.A.M. (1999). Plant regeneration through callus initiation from anther and ovules of *Scabiosa columbaria. Plant Cell Tiss. Org. Cult.*, **56:** 169-177.

Rosita, S.M.D. (1993). Effect of Boron and Cu on growth, yield and solasodin content of *Solanum khasianum* Clarke. *Pemberitian–Penelitian–Tanaman–Industri* (Indonesia), **15:** 110-114.

Rotino, G.L. (1996). Haploidy in egg plant. In: *In vitro production of haploids in higher plants*, Vol. 3, (eds.) Jain S.M. Sopory S.K. and Veillux, R.E., Kluwer Academic Publishers, Amsterdam, pp. 115-124.

Rotino, G.L., Falavigna, A. and Restaino, F. (1987). Production of anther-derived plantlets of egg plant. *Capsicum Newsletter*, **6:** 89-90.

Rotino, G.L. and Gleddie, S. (1990). Transformation of egg plant (*Solanum melongena* L.) using a binary *Agrobacterium tumefaciens* vector. *Plant Cell Rep.*, **9:** 26-29.

Rotino, G.L., Dorr, L., Pedrazzini, E. and Perri, E. (1995). Fusione di protoplastitra melanzana (*Solanum melongena* L.) *Solanum selvatici.* In: *Proc XXXIX Convegno Annuale S.I.G.A.* Vasto Marina, India, pp. 157.

Rotino, G.L., Perri, E., Acciarri, N., Sunseri, F. and Arpaa, S., (1997). Development of eggplant varietal resistance to insects and diseases via plant breeding. *Adv. Hort. Sci.*, **11:** 193-201.

Roy, S. K., Rahman, S.K.L. and Datta, P.C. (1988). *In vitro* propagation of *Mitragyna parviflora. Plant Cell Tiss. Org. Cult.*, **12:** 75-80.

Roy, S. K., Shariful Islam, M., Jayasree, S. and Hedinzzaman, S. (1996). In: *Plant Tissue Culture.* (Ed.) A.S. Islam, Oxford and IBH Publishing Co., pp. 8-15.

Roy, S.K., Islam, M.S., Sen, J., Hadiuzzaman, S. (1993). Propagation of flood tolerant Jack fruit (*Atrocarpus heterophyllus*) by *in vitro* culture. *Acta. Hort.*, **336:** 273-278.

Sadanandam, A. and Farooqui, M.A. (1991). Induction and selection of lincomycin-resistant plants in *Solanum melongena* L. *Plant Sci.*, **79:** 237-239.

Sahoo, Y. and Chand, P.K. (1998). *In vitro* multiplication of medicinal herb *Tridx procumbens* L. (maximum diasy, coat bottons). Influence of explanting season, growth regulator synergy, culture passage and planting substrate. *Phytomorphology,* **48:** 195-205.

Sahoo, Y., Pattnaik, S.K. and Chand, P.K. (1997). *In vitro* clonal propagation of an aromatic medicinal herb *Ocimum basilicum* L. (Sweet basil) by axillary shoot proliferation. *In vitro Cell Dev. Biol. Plant.,* **33:** 293-296.

Sahrawat, A.K. and Chand, S. (2001). Continuous somatic embyogenesis and plant regeneration from hypocotyl segments of *Psoralea corylifolia* Linn. An endangered and medicinally important Fabaceae plant.

Sakata, Y., Nishio, T. and Manma, S. (1989). Resistance of *Solanum* species to verticellium wilt and bacterial wilt In: EUCARPIA VII *the meeting on genetics and breeding on Capsicum and Egg plant,* 27-30 June, Karagujeva, Yugoslavia.

Samoylov, V.M. and Sink, K.C. (1996). Donor chromosome elimination and organelle composition of asymmetric somatic hybrid plants between an interspecific tomato hybrid and egg plant. *Theor. Appl. Genet.,* **93:** 268-274.

Sanders, P.R., Sammons, B., Kaniewski, W., Haley, L., Layton, J., Lavelle, B.J., Delannay, X. and Turner, N.E.(1992). Field resistance of transgenic tomatoes expressing the tobacomosaic virus coat protein gene. *Phytopathology,* **82:** 683-690.

Sanford, J.C. (1990). Biolistic plant transformation. *Physiol. Plant.* **79:** 206-209.

Sarmento, G.G., Alpert, K., Tang, F.A. and Punja, Z.K. (1992). *Plant Cell Tiss. Org. Cult.,* **31:** 185-193.

Sarvesh, A., Reddy, T.P. and Kavi Kishor, P.B. (1994). Androclonal variation in niger (*Guizotia abyssinica* Cass). *Euphytica,* **79:** 59-64.

Schank, R.V. and Hildebrandt, A.C. (1972). Medium and techniques for induction and growth of monocotyledonous and dicotyledonous plant cell cultures. *Can. J. Bot.*, **50:** 199-204.

Schery, R.W. (1952). *Plants for Man*, Prentice-Hall Inc., New York.

Schleiden, M.J. (1838). Beitrage zur phytogenesis Arch, Anat, Physiol, V. Wiss, Med (J. Muller), 137-176.

Schwann, T. (1839). Mikroskopische, Untersuchugen uberedie ubereinstimmung in der structure und dem wachstume der tiere and pflanzen leipzig, W. Engelmann, Nr. 76, Ostwalds, Klessiker der Exakten Wissenschaften, pp. 1910.

Seetharam, Y.N., Jyothishwaran,G., Sujeeth. H., Arvind Barad, Sharanabasappa, G. and Sangeetha Panchal, T. (2003). *Plant Cell Biotech. Mol. Biol.*, **4:** 17-22.

Sen, J. and Sharma, A.K. (1991). Micropropagation of *Withania somnifera* from germinating seeds and shoot tips. *Plant Cell Tiss. Org. Cult.*, **26:** 71-73.

Setterfield, G. (1963). Growth regulation in excised slices of Jerusalem artichoke tuber tissues. *Symp. Soc. Exp. Biol.*, **12:** 98.

Shah, S.D., Tobita, S. and Shono, M. (2003). Cation co-tolerance phenomenon in cell cultures of *Oryza sativa* adapted to Licl and Nacl. *Plant Cell Tiss. Org. Cult.*, **71:** 95-101.

Shahzad, A. and Siddique, A.S. (2000). *In vitro* organogenesis in *Ocimum sanctum* (L.)–A Multipurpose Herb. *Phytomorphology*, **50:** 27-35.

Shahzad, A., Hasan, H. and Siddiqui, A.S. (1999). Callus induction and regeneration in *Solanum nigrum* L. *in vitro*. *Phytomorphology*, **49:** 215-220.

Sharma, P. and Rajam, M.V. (1995). Genotype, explant and position effects on organogenesis and somatic embryogenesis in egg plant (*Solanum melongena* L.). *J. Exptl. Bot.*, **46:** 135-141.

Sharma, S.K. and Dhiman, R.C. (1998). *In vitro* clonal propagation of F_1 Hybrid of *Paulownia* (*P. fortunel xp. tomentosa*). *Phytomorphology*, **48:** 167-172.

Sharma, A.K., Mohanty, A., Singh, Y. and Tyagi, A.K. (1999). Transgenic plants for the production of edible vaccines and antibodies for immunotherapy. *Curr. Sci.*, **77:** 524-529.

Sharon, M. and D'Souza, M. (2000). *In vitro* clonal propagation of annatto (*Bixa orellana* L.). *Curr. Sci.,* **78:** 1532-1535.

Sheerman, S. and Bevan, M.W. (1988). A rapid transformation method for *Solanum tuberosum* using *Agrobacterium tumefaciens* vector. *Plant Cell Rep.,* **7:** 13-16.

Shen, B., Jensen, B.G. and Bohnert, H.J. (1997). Increased resistance to oxidative stress in transgenic plants by targeting mannitol biosynthesis to chloroplasts. *Plant Physiol.,* **113:** 1177-1183.

Shepherd, J.F., Bidney, D. and Shahin, E. (1980). Potato protoplasts in crop improvement. *Science,* **208:** 17-24.

Shivarajan, V.V. and Balachandran, I. (1999). *Ayurvedic drugs and their plant sources,* Oxford and IBH Publishing Co. Pvt. Ltd., 66, Janpath, New Delhi.

Sihachakr, D., Haicour, R., Serraf, I., Barriantos, E., Herbretean, C., Ducreaux, G., Rossignol, L. and Souvannavong, V. (1988). Electrofusion for the production of somatic hyrbid plants of *Solanum melongena* L. and *S. khasianum* C.B. Clarke. *Plant Sci.,* **57:** 215-223.

Singh, K.V. and Bansal, S.K. (2003). Larvicidal properties of a perennial herb *Solanum xanthocarpum* against vectors of malaria and dengue/DHF. *Curr. Sci.,* **84:** 749-751.

Sinha, R.K., Majumdar, K. and Sinha, S. (2000). Somatic embryogenesis and plantlet regeneration from leaf explants of *Sapindus mukorossi* Gaertn: A soap nut tree. *Curr. Sci.,* **78:** 620-623.

Sinnot, E.W. (1960). *Plant morphogenesis,* New York, McGraw-Hill.

Skoog, F. and Miller, C.O. (1957). Chemical regulation of growth and organ formation in plant tissue culture *in vitro. Symp. Soc. Exptl. Biol.,* **11:** 118-131.

Skriver, K. and Mundy (1990). Gene expression in response to abscisic acid and osmotic stress. *Plant Cell,* **2:** 503-512.

Steward, F.C., Mepes, M.G. and Mears, K. (1958). Growth and organized development of cultured cells organization in cultures grown from freely suspended cells. *Amer. J. Bot.,* **445:** 705-708.

Steward, F.C., Mapes, M.P., Kent, A.E. and Holbten, R.D. (1964). Growth and development of cultured plant cells. *Science,* **143:** 20-27.

Stoehr, M.V. and Zsuffa, L. (1990). Induction of haploids in *Populus maximowiczii* via embryogenic callus. *Plant Cell Tiss. Org. Cult.,* **23:** 49-53.

Subhash, K. and Christopher, T. (1988). Direct plantlet formation in cotyledon culture of *Capsicum frutescens. Curr. Sci.,* **57:** 99-100.

Subhash, K., Venkataiah, P. and Bhaskar, P. (1996). Induction of streptomycin-resistant plantlets in *Capsicum annum* L. through mutagenesis *in vitro. Plant Cell Rep.,* **16:** 111-113.

Subhashini, K. and Reddy, G.M. (1990). Effect of salt stress on enzyme activities in callus cultures of tolerant and susceptible rice cultivars. *Indian J. Exptl. Biol.,* **28:** 277-279.

Sudershan, C. (1998). Shoot bud regeneration from leaf explants of a medicinal plant *Enicostemma axillare. Curr. Sci.,* **74:** 1099-1100.

Sudershan, C., Aboel, M. N. and Hussain, J. (2000). *In vitro* propagation of *Zizyphus mauritiana* cultivar umrdn by shoot tip and nodal multiplication. *Curr. Sci.,* **80:** 290-292.

Sudha, G.C. and Seeni, S. (1994). *In vitro* propagation and field establishment of *Adhatoda beddomei* C.B. Clarke, a rare medicinal plant. *Plant Cell Rep.,* **13:** 203-207.

Sunitha, C. and Handique, P.J. (2000). High frequency *in vitro* shoot multiplication of *Plumbago indica* a rare medicinal plant. *Curr. Sci.,* **78:** 1187-1188.

Svab, Z. and Maliga, P. (1986). *Nicotina tabacum* mutants, with chloroplast encoded streptomycin resistance and pigment deficiency. *Theor. Appl. Genet.,* **72:** 637-643.

Svab, Z., Harper, E.C., Jones, J.D.G. and Maliger, P. (1990). Aminoglycoside-3"-adenyltransferase, confers resistance to spectinomycin and streptomycin in *Nicotiana tabacum. Plant Mol. Biol.,* **14:** 197-205.

Swanson, E.B. and Erickson, L.R. (1989). Haploid transformation in *Brassica napus* using an octopine producing strain of *Agrobacterium tumefaciens. Theor. Appl. Genet.,* **78:** 831-835.

Swanson, E.B., Coumans, M.P., Brown, G.L., Patel, J.D. and Beverdorf, W.D. (1988). The characterization of herbicide tolerant plants in *Brassica napus* L. after *in vitro* selection of microspores and protoplasts. *Plant Cell Rep.,* **7:** 83-87.

Swanson, E.B., Harrgesell, M.J., Arnoldo, M., Sippell, D.W. and Wang, R.S.C. (1989). Microspores mutagenesis and selection of canola plants with field tolerance to the imidazolines. *Theor. Appl. Genet.,* **78:** 525-530.

Tabone, T.J., Felker, P., Bingham, R.L., Rayes, I. and Loughrey, S. (1986). Techniques in the shoot multiplication of the Leguminous tree *Prosopis alba* clone B2 V50 for *Ecolmanay,* **V16:** 191-200.

Takaba, I., Labib, G. and Melchers, G. (1971). Regeneration of whole plants from isolated mesophyll protoplasts of tobacco. *Naturwissen,* **58:** 318-320.

Tal, M. (1994). *In vitro* selection for salt-tolerance in crop plants–theoretical and practical considerations. *In vitro Cell Dev. Biol. Plant,* **34:** 175-180.

Tan, M.M.C., Colijn-Hooymans, C.M., Lindhout, W.H. and Kool, A.J. (1987). A comparison of shoot regeneration from protoplasts and leaf discs of different genotypes of the cultivated tomato. *Theor. Appl. Genet.,* **75:** 105-108.

Taylor, D.C., Ferrie, A.M.R., Keller, W.A., Giblin, E.M., Pass, E.W. and MacKenzie, S.L. (1993). Bioassembly of acyl lipids in microspore derived embryos of *Brassica campestris* L. *Plant Cell Rep.,* **12:** 375-284.

Tejavathi, D.H. and Bhuvana, B. (1998). *In vitro* morphogenetic studies in *Solanum viarum* Dunal. *J. Swamy Bot. Cl.,* **15:** 27-30.

Thomas, E. and Street, H.E. (1970). Organogenesis in cell suspension cultures of *Atropa belladonna* L. and *Atropa balladonna* Cv. Lutea Doll. *Ann. Bot.,* **34:** 657-669.

Thomas, E., Bright, S.W.J., Franklin, J., Lancaster, V. and Mi Flia, B.J. (1982). Variation amongst protoplasts derived from potato (*Solanum tuberosum*). *Theor. Appl. Genet.,* **61:** 65-68.

Thomas, J.C., Arnold, R.L. and Bohnert, H.J. (1992). Influence of NaCl on growth, proline, phosphoenol pyruvate carboxylase levels in *Mesembryanthemum crystallinum* suspension cultures. *Plant Physiol.*, **98:** 626-631.

Thorpe, T.A. (1980). Organogenesis *in vitro*: structural, physiological and biochemical aspects. *Int. Rev. Cytos. Suppl. IIA*, 71-112.

Tisserat, B. (1987). *Embryogenesis, organogenesis and plant regeneration in plant cell culture: A practical approach.* (ed.) Dixon, R.A., IRL Press, Oxford.

Tiwari, A.S., Minakshi, S. and Ranga Swamy, N.S. (1999). Somatic embryogenesis and plant regeneration in three cultivars of white jute. *Phytomorphology.*, **49:** 303-308.

Tiwari, V., Tiwari, K.N. and Singh, B.D. (2000). Suitability of liquid culture for *in vitro* multiplication of *Bacopa monniera* (L.) Wettst. *Phytomorphology*, **50:** 337-342.

To, K.Y., Chem, C.C. and Lai, Y.K. (1989). Isolation and characterization of streptomycin-resistant mutants in *Nicotiana plumbaginifolia. Theor. Appl. Genet.*, **78:** 81-86.

To, K.Y., Lai, Y.K., Feng, T.Y. and Chem, C.C. (1992). Restriction endonuclease analysis of chloroplast DNA from streptomycin resistant mutants of *Nicotiana plunbaginifolia. Genome*, **35:** 220-224.

Tricoli, D.M., Kim, J., Carney, Russell, P.F., Russell, M.M., Groff, D.W., Hadden, K.C., Himmel, P.T., Hubbard, J.P., Boeschore, M.L. and Quemada, H.D. (1995). Field evaluation of transgenic squash containing single or multiple virus coat protein gene constructs for resistance to cucumber mosaic virus. *Biotechnology*, **13:** 1458-1465.

Umbek, P., Johnson, G., Barton, K. and Swain, W. (1987). Genetically transformed cotton (*Gossypium hirsutum* L.) plants. *Biotechnology*, **5:** 263-266.

Vaek, N., Reynaerts, A., Hoffe, H., Jansens, S., Beukeleer, M.D., Dean, C., Zabeau, M., Montagu, M.C. and Leemans, J. (1987). Transgenic plants protected from insect attack. *Nature*, **328:** 33-37.

Van Overbeek, J., Conklin, M.E. and Blakslec, A.F. (1941). Factors in coconut milk essential for growth and development of very young *Datura* embryos. *Science*, **94:** 350-351.

Van Rockel, J.C., Damm, B., Melchers, L.S. and Hoekema, A. (1993). *Plant Cell Rep.*, **12:** 644-647.

Van Swaaij, A.C., Jacobsen, E., Kiel, J.A.K.W. and Feenstra, W.J. (1986). Selection, characterization and regeneration of hydroxyproline resistant cell lines of *Solanum tuberosum* tolerance to NaCl and freezing stress. *Physiol. Plant,* **68:** 359-366.

Vanden Elzen, P.J.M. *et al.* (1993). Virus and Fungal resistance: From laboratory to field. *Philos Trans Roy Soc. London. Ser. B.,* **342:** 271-278.

Varisaimohamed, S., Jowar, M. and Jayabalan, N. (1998). Effects of Ads, BAP and IBA on plant regeneration from *Macrotyloma uniflorum* (lam.) Verde: *Phymorphology,* **48:** 61-65.

Vasil, V. and Hildbrandt, A.C. (1965). Differentiation of tobacco plants from single isolated cells in microculture. *Science,* **150:** 889–892.

Vasil, I.K. and Nitsch, C. (1975). Experimental production of pollen haploid and their uses. *Pflanzen Physiol.,* **76:** 191-212.

Vasil, I.K., Ahuja, M.R. and Vasil, V. (1979). Plant tissue culture in genetics and plant breeding. *Adv. Genet.,* **20:** 127-215.

Vasil, V. and Vasil, I.K. (1980). Isolation and culture of cereal protoplasts part 2 embryogenesis and plantlet formation from protoplasts of *Pennisetum americanum. Theor. Appl. Genet.,* **56:** 97-99.

Vasil, I.K., Vasil, V., Liu, C., Ozias-Akins, P., Haudu, Z. and Wang, Y. (1982). Somatic embryogenesis in cereals and grasses. In: *Variability in plant regenerated from tissue cultures* (eds.) Earle, E.D. and Demarly, Y., Preger, New York.

Vasil, V. and Vasil, I.K. (1984). In: *Cell culture and somatic cell genetics of plants.* Vasil I.K. (ed.) Vol. 1, Academic Press, Orlando, pp. 36-41.

Venkatachalam, P., Sankara Rao, K., Kavi Kishor, P.B. and Jayabalan, N. (1998). Regeneration of late leaf spot–resistnat ground nut plants from *Cercosporidium personatum* culture, filtrate-treated callus. *Curr. Sci.,* **74:** 1-10.

Venkatachalam, P., Geetha, N., Khandelwal, A., Shaila, M.S. and Lakshmi Sita, G. (2000). *Agrobacterium* mediated genetic transformation and regeneration of transgenic plants from cotyledon explants of groundnut (*Arachis hypogaea* L.) via somatic embryogenesis. *Curr. Sci.*, **78:** 1130-1136.

Venkataiah, P. (2000). *In vitro* studies on *Capsicum*, Ph.D. thesis, Kakatiya University, Warangal (A.P.), India.

Venkataiah, P. and Subhash, K. (2001). Genotype, explant and medium effects on adventitious shoot bud formation and plant regeneration in *Capsicum annuum* L. *J. Genet. and Breed.*, **55:** 143-149.

Verbruggen, N., Villarroel, R. and Montagu, M.V. (1993). Osmoregulation of a pyrroline-5-carboxylate reductase gene in *Arabidopsis thaliana. Plant Physiol.*, **103:** 771-781.

Vernon, D.M. and Bohnert, H.J. (1992). A novel methyl transferase induced by osmotic stress in the facultative halophyte, mesembryanthemum crystallinum. *EMBO Journal*, **11:** 2077-2985.

Vincent, K.A., Mathew, K.M. and Hariharan, M. (1992). Micropropagation of *Kaempfera galanga* (L.). A Medicinal plant. *Plant Cell Tiss. Org. Cult.*, **28:** 229-230.

Virendra, K., Gautam, Nanda, K. and Gupta, S.C. (1993). Development of shoots and roots in anther-derived callus of *Azadirachta indica*. Juss–a medicinal tree. *Plant Cell Tiss. Org. Cult.*, **34:** 13-18.

Wagly, L.M., Gladfelter, H.J. and Phillips, G.C. (1987). *De novo* shoot organogenesis of *Pinus eldarica* Medw., *in vitro* macro and micro photographic evidence of *de novo* regeneration". *Plant Cell Rep.*, **6:** 167-171.

Waller, G.R. (1987). In: *Allelochemicals: Role of agriculture and forestry. Amer. Chem. Soc.* Washington DC (ACS Symp Ser 330).

Wang, P.J. and Hu, N.Y. (1982). *In vitro* mass tuberization and virus free potato production in Taiwan. *Amer. Potato. J.*, **59:** 33-39.

Weretilnyk, E.A. and Hanson, A.D. (1990). Molecular cloning of a plant betaine-aldehyde dehydrogenase an enzyme implicated in adaptation to salinity and drought. *Proc. Natl. Acad. Sci., USA*, **87:** 2745-2749.

White, P.R. (1939). Potentially unlimited growth of excised plant callus in artificial medium. *Amer. J. Bot.*, **26:** 59-64.

Widholm, J.M. (1974). Cultured carrot cell mutants: 5-methyl tryptophan resistance trait carried from cell to plant and back. *Plant Sci. Lett.*, **3:** 322-330.

Wilkins, C.P. and Dodds, J.H. (1983). The application of tissue culture to plant genetic conservation, *Science Progress*, **68:** 259-284.

Withers, L.A. (1989). *In vitro* conservation and germplasm utilisation. In: T*he use of plant genetic resources*, (eds.) A.H.D. Bronons, D.H. Frankel, D.R. Marshall and J.J. Williams. Cambridge University Press, Cambridge, pp. 309-334.

Yadav, J.S. and Rajam, M.V. (1997). Spatial distribution of free and conjugated polyamines in leaves of *Solanum melongena* L. associated with differential morphogenetic capacity: efficient somatic embryogenesis with putresine. *J. Exptl. Bot.*, **48:** 1537-1547.

Yadav, J.S. and Rajam, M.V. (1998). Temporal regulation of somatic embryogenesis by adjusting cellular polyamine content in egg plant. *Plant Physiol.*, **116:** 617-625.

Yadav, V., Malan, L. and Jaiswal, V.S. (1990). Micropropagation of *Morus niger* from shoot tip and nodal explants of mature trees. *Scientia Hort.*, **44:** 61-64.

Yamamoto, Y., Mizuguchi, R. and Yamada, Y. (1982). Selection of high and stable pigment producing strain in culture *Euphorbia* Mill. *Theor. Appl. Genet.*, **61:** 113-116.

Yeoman, M.M. and Macleod, A.J. (1977). Tissue (callus) cultures techniques In: *Plant Tissue and Cell Culture*, 2nd edt (ed.) H.E. Street, Oxford Black Well Scientific Publications, pp. 31-59,

Zaman, A., Islam, R., Joarder, O.I. and Barman, A.C. (1996). Clonal propagation of mulberry plants through, *in vitro* techniques. In: *Plant Tissue Culture*, (ed.) A.S. Islam Oxford and IBH publishing Co., pp. 71-75.

Zhou, H. and Konzak, C.F. (1992). Genetic control of green plant regeneration from anther culture of wheat. *Genome*, **35:** 957-961.

Zimmerman (1993). Somatic embryogenesis: A model for early development in higher plants. *J. Plant Cell*, **5:** 141-143.

Index

L

M

N

O

P

www.ingramcontent.com/pod-product-compliance
Ingram Content Group UK Ltd.
Pitfield, Milton Keynes, MK11 3LW, UK
UKHW021007290726
14059UKWH00001BA/2

9 789351 306375